Essential Python for the Physicist

Giovanni Moruzzi

Essential Python for the Physicist

Springer

Giovanni Moruzzi
Department of Physics "Enrico Fermi"
University of Pisa
Pisa, Italy

ISBN 978-3-030-45029-8 ISBN 978-3-030-45027-4 (eBook)
https://doi.org/10.1007/978-3-030-45027-4

This Springer imprint is published by the registered company Springer Nature Switzerland AG
The registered company address is: Gewerbestrasse 11, 6330 Cham, Switzerland

Preface

Python is a *high-level*, general-purpose programming language created by Guido van Rossum and first released in 1991. Python is an *interpreted* language emphasizing code readability: Its syntax allows programmers to express concepts in fewer lines of code than other languages such as C++ or Java. Python features a dynamic type system and automatic memory management. It supports multiple programming paradigms, including object-oriented, imperative, functional and procedural, and has a very large and comprehensive standard library.

The purpose of the present book is to introduce the physics student, or, more generally, a natural science student, to Python. The book requires no previous experience with other programming languages even if, of course, such a previous experience, if present, makes learning Python easier and faster. We start with a brief introduction to the basics of Python programming in interactive mode, then present scripts, or programs, of increasing complexity, and finally reach fundamental computer animation. Special regard is given to simple animations of physical interest. In fact, a great advantage of Python is its being, at least at the moment, the only programming language allowing cross-platform graphics. This means that your programs involving graphics and animation will run on Linux, macOS, and Windows with no need of changes.

We assume some previous knowledge of classical physics and calculus. Quantum physics is required only for reading Sect. 5.5, which can be skipped by the uninterested reader. Analogously, Subsections 7.6.1 and 7.6.2 can be skipped by readers not interested in the details of the numerical integration of the ordinary differential equations, but only in solving them with the available built-in functions provided by Python.

Covering the whole Python language would require several volumes, and is definitely outside the scope of the present book. However, after reading this book, one should acquire feelings and ideas on where to look, and what to look for, on the internet, for tackling more specialized tasks involving Python packages and built-in functions not discussed here.

Although we provide brief discussions of all numerical methods involved in the programs presented in this book, this is not a book on numerical methods. Many good books on numerical methods are available to the interested reader. In any case, Python provides a very extensive collection of precompiled built-in numerical functions in its libraries, which allow you to tackle most physical problems.

The author is indebted to Bruce Sherwood for valuable comments and suggestions, to Giuseppe Bertin and Andrea Macchi for suggesting interesting exercises.

Pisa, Italy
February 2020

Giovanni Moruzzi

Contents

Chapter 1
Python Basics and the Interactive Mode

1.1 Using the Command Line

When you start writing your own computer programs it can be useful, even if not mandatory, to interact with your computer via the *command line*. Since nowadays the overwhelming majority of the computer users ignore even the existence of the command line, we discuss it briefly in this section.

In order to use the command line you start by opening a window, sometimes called *terminal*, *shell* or *console*, in the computer monitor. The terminal contains a *command prompt* comprising a sequence of characters indicating readiness to accept commands. The actual prompt is different from one operating system to another, and some operating systems allow you to customize it. In the rest of this book the prompt will be represented by the two-character sequence `$>`. The command prompt literally prompts the user to take action. Figure 1.1 shows a typical terminal, and its prompt, on Ubuntu Linux. Terminals of other operating systems look similar. A simple way to open a terminal under Linux is pressing `Ctrl+Alt+T` (pressing the keys `Ctrl`, `Alt` and `T` simultaneously on the keyboard).

If you are using Mac OS X, the `Terminal` app is in the `Utilities` folder in `Applications`. To open it, either open your `Applications` folder, then open `Utilities` and double-click on `Terminal`, or press `Command-spacebar` to launch `Spotlight` and type "Terminal," then double-click the search result.

Under Windows you can open a terminal by clicking the `Start` button, typing `cmd` and pressing the <Enter> key.

The command line on a terminal was the primary means of interaction with most computer systems in the mid-1960s. In those times the "terminal" initially consisted of a teleprinter, later replaced by a keyboard and cathode-ray monitor. The command

Electronic supplementary material The online version of this chapter (https://doi.org/10.1007/978-3-030-45027-4_1) contains supplementary material, which is available to authorized users.

G. Moruzzi, *Essential Python for the Physicist*,
https://doi.org/10.1007/978-3-030-45027-4_1

Fig. 1.1 A Linux terminal. Here the prompt is the sequence `giovanni@moruzzi1: ~>`

line continued to be used throughout the 1970s and 1980s on personal computer systems including MS-DOS, CP/M and Apple DOS, the "terminal" being replaced by a "terminal emulator", a window on the computer monitor where you could type your commands. The interface between your commands and the computer actions is usually implemented with a *command line shell*, a program that accepts commands as text input and converts commands into appropriate operating system functions.

Once you have opened a terminal you can start typing commands, hitting <Enter> at the end of each. Each operating system has its own list of native commands, and you can add your personal commands. For instance, if you type "`ls -l`" in a Linux or Mac terminal, you will get the list of the contents of the current directory. The same result is obtained by typing "`dir`" in a Windows terminal. In this context you don't need to learn the whole lists of available commands for your operating system: when the command line is needed, we shall tell you what to type.

1.2 Installing Python

1.2.1 General

Obviously, in order use Python you must have Python (we shall use the Python 3 version) installed in your computer. The Ubuntu and the Debian distributions of Linux come with both Python 2 and Python 3 already installed by default, thus you can skip the following Sects. 1.2.2 and 1.2.3 if you use Ubuntu or Debian. If you have Windows or macOS (previously Mac OS X and later OS X) you will probably need to install Python 3. In this case Sects. 1.2.2 and 1.2.3 tell you how to do it.

However, even if you are a Linux user, you might be interested in using Python in an *integrated development environment* (IDE) rather than through the command line in a terminal (particularly if Sect. 1.1 has scared you!) An IDE is a software application that provides comprehensive facilities to computer programmers for software development. An IDE normally consists of a source-code editor for typing your program code, build automation tools, and a debugger. By *build automation* we mean the combined processes of compiling computer source code into binary code, packaging binary code, and running automated tests. A very good option for Python is Anaconda, a free and open-source distribution of the Python and R programming languages for scientific computing. Anaconda is available for Linux, Windows and macOS, and you can easily download it from their site

`https://www.anaconda.com/distribution/`

whatever your operating system. Choose the Python 3 version for your operating system, and follow the download instructions on your browser. Installing Anaconda automatically installs also a version of Python, thus, if Anaconda is your choice, you can skip Sects. 1.2.2 and 1.2.3 even if you are a Windows or Mac user. Once Anaconda is installed, launch Spyder 3, and you obtain the window shown in Fig. 1.2. The Spyder window is divided into three rectangular subwindows: the left subwindow is an editor for typing programs, or *scripts*, see Chap. 2; the lower-right subwindow is a console where you can use Python interactively, as discussed in this chapter starting from Sect. 1.3. An alternative good IDE is IDLE (Integrated Development and Learning Environment), which is also available for Windows, Linux and macOS. However, in the present book we shall discuss only Spyder and the command-line terminal

1.2.2 Downloading Python for Windows

1.2.2.1 Step 1: Download the Python 3 Installer

Open a browser window and navigate to the Download page for Windows at python.org. Underneath the heading at the top that says Python Releases for Windows, click on the link for the Latest Python Release–Python 3.x.x. Scroll to the bottom and select either Windows x86-64 executable installer for 64-bit or Windows x86 executable installer for 32-bit.

1.2.2.2 Step 2: Run the Installer

Once you have chosen and downloaded the installer of your choice, simply run it by double-clicking on the downloaded file. You should see a dialog similar to Fig. 1.3 on your computer monitor.

Then just click `Install Now`. This will download Python 3, the *pip* Python package manager and Python documentation. That should be all there is to do. A few minutes later you should have a working Python 3 installation on your Windows system.

1.2.3 Downloading Python for macOS

In the following, the symbol `$>` stands for the command prompt on the terminal, while a backslash (\) will mean that a long single command, that actually you must type in a single line, has been split into two lines to fit the page.

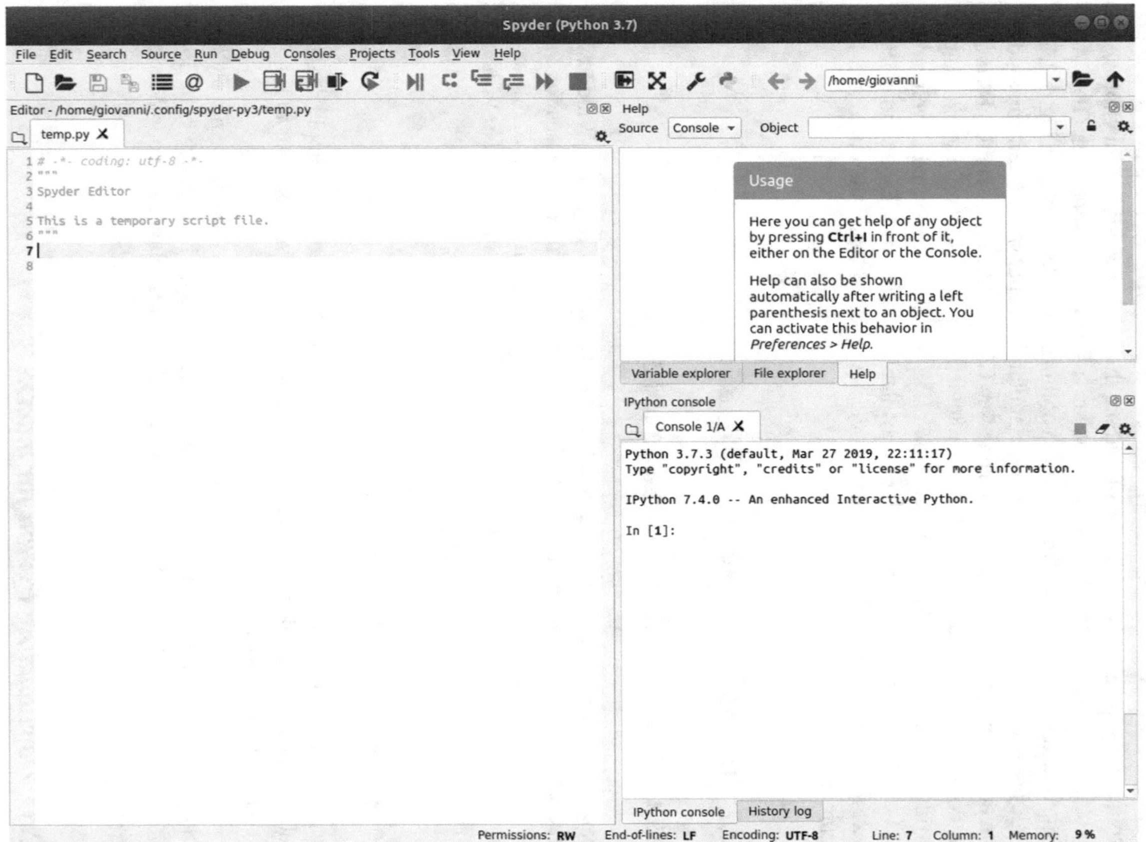

Fig. 1.2 The Spyder3 integrated development environment

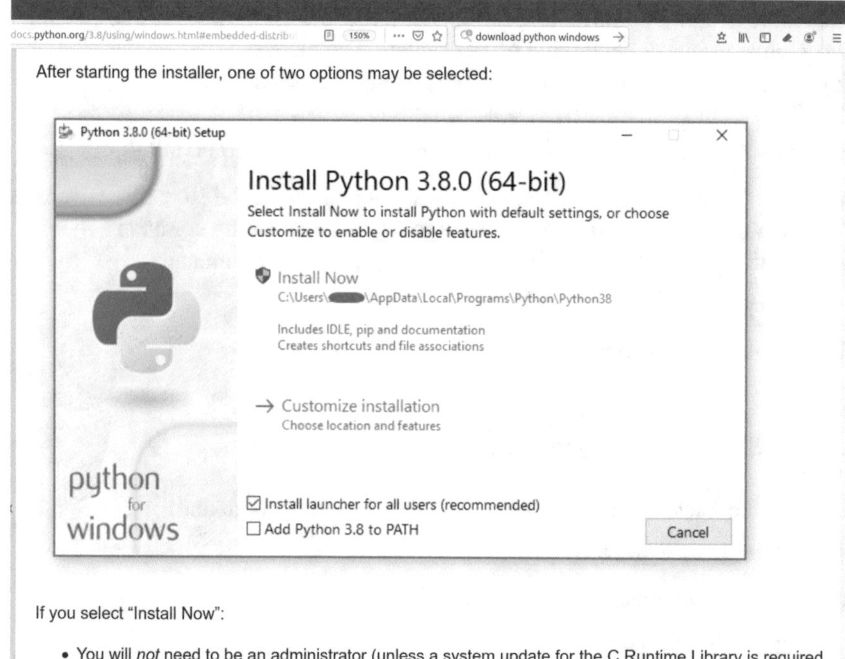

Fig. 1.3 Python installer for Windows

1.2.3.1 Step 1: Confirm your Python version

Although Python 2 is installed by default on Apple computers, Python 3 is not. You can confirm this by typing in Terminal

```
$> python --version
Python 2.7.15
```

To check if Python 3 is already installed try running the command

```
$>python3 --version.
```

Most likely you will see an error message, but it is worth checking. Even if you have a version of Python 3, we want to be on the most recent release, which is 3.7.0 at this point in 2018.

1.2.3.2 Step 2: Install Xcode and Homebrew

It is advisable to use the package manager Homebrew to install Python 3. Homebrew depends on Apples Xcode package, so run the following command to install it

```
$> xcode-select --install
```

Click through all the confirmation commands (Xcode is a large program so this might take a while to install depending on your internet connection).
Next, install Homebrew with the following (long) command:

```
/usr/bin/ruby -e "\$(curl␣-fsSL␣\
https://raw.githubusercontent.com/Homebrew/install/master/install)"
```

Note: You can also find this command on the homepage of the Homebrew website. It is easier to copy and paste rather than typing, since its a long command.

To confirm that Homebrew installed correctly, run this command:

```
$> brew doctor
```

Your system is ready to brew.

1.2.3.3 Step 3: Install Python 3

To install the latest version of Python, run the following command:

```
$> brew install python3
```

Now let's confirm which version was installed:

```
$> python3 --version
Python 3.7.0
```

To open a Python 3 shell from the command line type `python3`:

```
$> python3
Python 3.7.0 (default, Jun 29 2018, 20:13:13)
{[Clang 9.1.0 (clang-902.0.39.2)]} on darwin
Type "help", "copyright", "credits" or "license" for more information.
>>>
```

When you want to exit, type `exit()` and then Return, or `Ctrl-D` (press the Control and D keys at the same time).

1.3 Using Python in Interactive Mode

Once you have installed Python in your computer, or a friend has installed it for you, you have the two possibilities discussed above:

1. You can open a terminal and access Python through the command line, provided that Sect. 1.1 did not scare you.
2. You can start using Python through an IDE, here we shall consider Spyder.

Python programs, or Python commands, are executed by an interpreter. There are two ways of using Python: (i) the *interactive mode*, which we discuss in the rest of the present chapter, and (ii) Python programs (also called *Python scripts*), which we shall discuss in the rest of the book.

When working in the interactive mode you type Python commands one by one (or small groups of commands, as we shall see below) at the Python prompts, and Python immediately interprets what you wrote and executes your commands. This is what you must do according to your choice between command-line terminal or Spyder IDE,

1. If you choose the command line in a terminal, you start the Python interactive mode by simply typing `python3`, and pressing `<Enter>`, at the command prompt. Immediately after entering the interactive mode you see something like this on your monitor

```
$> python3
Python 3.5.2 (default, Nov 17 2016, 17:05:23)
[GCC 5.4.0 20160609] on linux
Type "help", "copyright", "credits" or "license" for more\
information.
>>>
```

   Obviously some of what you see above will change according the Python version installed in your computer and to your operating system. The back slash (\) at the end of the fourth line means that the whole line was too long to fit in the page of the book, so it was split and continued below. The symbol »> in the last line is the *Python prompt*, prompting you to enter your first Python command.
2. Alternatively, if you choose Anaconda and Spyder, the interactive mode is available in the bottom-right subwindow of Fig. 1.2, called the *IPython* console (IPython stands for "Interactive Python"). You type your first Python command at the prompt "`In [1]:`".

The advantage of the interactive mode is that you can immediately see how Python reacts to your commands, and discover possible errors immediately. On the other hand, the interactive mode becomes uncomfortable when you write long codes, that are better handled by Python *scripts*, to be introduced in Chap. 2.

This is what you see if you type, for instance, "`print('Hello World!!!')`" at the Python prompt in the terminal

```
>>> print('Hello World!!!')
Hello World!!!
>>>
```

while this is what you see in the IPython console of Spyder

```
In [1]: print('Hello World!!!')
Hello World!!!

In [2]:
```

`print()` is a command (actually, a *function*, to be discussed in Sect. 2.2) that tells Python to print the content of the parentheses (the *argument* of the function). The single quotes (`'`) tell Python to interpret their content as a sequence of printable characters (a *string*) to be printed as they are, not as a *variable* (see Sect. 1.4). A string can also be delimited by double quotes, `"also this is a string"`:

single quotes and double quotes are fully equivalent in Python. The final prompts, >>> in the terminal, and `In [2]:` in the IPython console, tell you that Python is waiting for your next command. In the examples of the rest of this chapter we shall show only the terminal prompt ">>>", if you use Spyder this will be replaced by the prompt "`In [n]`", where `n` is a progressive natural number.

Python in interactive mode can be used as a desktop calculator, for instance:

```
>>> 15+16
31
>>>_+9
40
>>>
```

the underscore (_) in the command `_+9` means that 9 must be added to the previous result. This works only in interactive mode, not in scripts.

1.4 Variables

1.4.1 Variable Types

In computer programming, a variable is a memory storage location associated to a symbolic name (its *identifier*), which contains some quantity of information (the variable *value*). Differently from other programming languages, Python variables do not need explicit declaration to reserve memory space. The memory allocation (or *variable declaration*) occurs automatically the first time the variable appears at the left of an equal sign (=), which serves as *assignment operator*. In other words, the equal sign assigns values from its right side to the variable at its left side. When choosing the name for a new variable remember that

1. The name of a variable must begin with a letter (a–z, A–Z) or underscore (_).
2. The following characters of the name can be letters, numbers or underscores.
3. Variable names are *case sensitive*, for instance, `cat`, `Cat` and `CAt` are three different variables.
4. Variable names can have any (reasonable) length.
5. There are reserved words, or *keywords*, used by Python to define the syntax and structure of the Python language. You cannot use keywords as variable names.

The first time a variable is used, it must appear at the left side of an assignment operation. The assignment first reserves space for the variable in the computer memory, then copies what is at the right side of the = sign into the variable storage location. Successive assignments involving the same variable only change its previous value, keeping its memory location.

Python variables belong to five standard data types:

1. Numbers
2. Strings

3. Lists
4. Tuples
5. Dictionaries

Numbers can be *integers* like `10`, *long integers* like `51924361L`, *floats* like `132.57`, and *complex* like `13.2+2.5j` (`j` standing for the imaginary unit). We have already met strings in Sect. 1.3: strings are sequences of characters enclosed in quotation marks. As mentioned above, both pairs of single quotes and pairs of double quotes are allowed: `'platypus'` and `"platypus"` are equivalent. Lists and tuples are discussed in Sect. 1.10, dictionaries in Sect. 1.14. In this section we handle strings and numbers. For example you can type

```
>>> counter=100          # An integer assignment
>>> mass=10.0            # A floating point
>>> velocity=15.22       # A floating point
>>> name="John"          # A string

>>> print(name)
John
>>> print(mass*velocity)
152.20000000000002
>>> print(format(mass*velocity,"10.3f"))
152.200
>>> momentum=mass*velocity
>>> print(momentum)
152.20000000000002
```

You have to type only what follows the Python interpreter prompts (>>>), while all the lines not starting with prompts are printed by Python automatically. Note the presence of *rounding errors*. This is due to the fact that all numbers, integers or float, are stored in binary form in a computer memory. While this does not give problems with integers, you must remember that binary fractions only terminate if the denominator has 2 as the only prime factor. Thus, most rational numbers (and all irrational numbers) need an infinite number of bits for an exact binary representation. However, obviously, only a finite number of bits is available for storing a variable, and this leads to rounding errors. This is the reason for the apparently strange value of the product `mass*velocity`. Often a *formatted output* leads to a more "aesthetic" result (see Sect. 1.15). Here, the command `print(format(mass*velocity,"10.3f"))` tells Python to print the result as a 10 *character* number, with 3 digits after the decimal point. The 10 *characters* include all digits before and after the decimal point, leading blanks, the decimal point itself and, in the case of a negative number, the minus sign (see Sect. 1.15).

As in most other programming languages the asterisk, or star, sign (`*`) is used for multiplication (see Sect. 1.5). In Python, the hash symbol (#) and everything that follows it in a same line are considered a comment, and are ignored in the program execution. A multiline comment is delimited by triple quotes

```
'''
this is a multiline
comment
'''
```

also triple double quotes """ work as comment delimiters.

1.4.2 Variable-Type Conversion

Sometimes it is necessary to perform conversions between the built-in variable types, in order to manipulate values in a different way. To convert between variable types you simply use the type name as a function. In addition, Python provides several built-in functions to perform special kinds of conversions. All of these functions return a new object representing the converted value. The most relevant examples follow.

1.4.2.1 Conversions Between Integers and Floats

When you type a sequence of digits not containing a decimal point, Python interprets it as an integer. This will be often what you wish, but you might need your number stored in memory as a float. The conversion is achieved through the function `float()`, this is how it works

```
>>> x=35
>>> xf=float(x)
>>> print(x,xf)
35 35.0
```

here the variable `x` is stored in memory as an integer, `xf` as a float. Conversion from float to integer is achieved through the function `int()`

```
>>> y=48.9
>>> yi=int(y)
>>> print(y,yi)
48.9 48
```

note that this conversion does not round `y` to the nearest integer: the function `int()` simply cuts off the decimal point and the following digits. If what you want is *rounding to the nearest integer*, you can simply add 0.5 to the float

```
>>> yr=int(y+0.5)
>>> print(yr)
49
```

Remember that you must add −0.5 if the float to be rounded is negative.

```
>>> y=-15.8
>>> yr1=int(y+0.5)
>>> yr2=int(y-0.5)
>>> print(y,yr1,yr2)
-15.8 -15 -16
```

1.4.2.2 Conversions Between Numbers and Strings

A string is a sequence of characters. Thus, for instance, the two consecutive characters `35` might be stored in the computer memory as an integer, as a float, or as a string comprising the two characters `'3'` and `'5'`. As we have seen above, the command `x=35` assigns the value 35 to the integer variable `x`. This can be converted to a string variable, say `xs`, by the `str()` function:

```
>>> x=35
>>> xs=str(x)
>>> xs
'35'
>>> print(xs)
35
```

note that, when you simply type `xs` in Python interactive mode, Python prints the string in quotes, while the string is printed without quotes by the command `print(xs)`. However, see the effect of the format type `r` on the `print()` function in Sect. 1.15.3. Converting an integer to a string can be useful, for instance, for counting its digits through the `len()` function

```
>>> x=23457439
>>> xs=str(x)
>>> print(len(xs))
8
```

Conversion of strings to integers is done by the `int()` function, from string to floats by the `float()` function, for instance

```
>>> xs='542'
>>> xi=int(xs)
>>> xf=float(xs)
>>> print(xs,xi,xf)
542 542 542.0
```

If a string contains a character that is not a digit, a decimal point, a leading minus or plus sign, or leading and/or trailing blanks, it cannot be converted into a number, and Python will report an error

```
>>> xs='543f'
>>> x=int(xs)
Traceback (most recent call last):
File "<stdin>", line 1, in <module>
ValueError: invalid literal for int() with base 10: '543f'
```

Here, and in following Python error messages, `File "<stdin>"` means that the error was found in the *standard input*, i.e., in what you typed from the computer keyboard. Analogously, `"<stdout>"` stands for *standard output* , which is the computer monitor.

1.5 Arithmetic Operators

The basic arithmetic operators that can be applied to numeric variables are

$+$ addition	$2.5+3.0=5.5$	$-$ subtraction	$2.5-3.0=-0.5$
$*$ multiplication	$2.5*3.0=7.5$	$/$ division	$2.5/2=1.25$
$//$ floor division	$7//2=3;\ -7//2=-4$	$\%$ modulus	$11\%3=2$
$**$ exponentiation	$11**2=121$		

Note that the *floor* of -3.5 is -4, since $-4<-3.5$.

1.6 Assignment Operators

Apart from the already discussed = operator, other assignment operators are obtained by combining the basic arithmetic operators with the = operator as follows

`+=`	`c+=a` is equivalent to `c=c+a`	`-=`	`c-=a` is equivalent to `c=c-a`
`*=`	`c*=a` is equivalent to `c=c*a`	`/=`	`c/=a` is equivalent to `c=c/a`
`%=`	`c%=a` is equivalent to `c=c%a`	`**=`	`c**=a` is equivalent to `c=c**a`
`//=`	`c//=a` is equivalent to `c=c//a`		

For instance, you can type

```
>>> a=8
>>> b=3
>>> b+=a
>>> print(a,b)
8 11
>>>
```

1.7 Comparison and Logical Operators

The comparison operators are

$>$ greater than	$<$ less than	$==$ equal to
$!=$ not equal to	$>=$ greater than or equal to	$<=$ less than or equal to.

The result of a comparison is a Boolean value, denoted by either *True* or *False* in Python. For instance you can type

```
>>> print('10>15 is ',10>15)
10>15 is  False
```

Note that here the `print()` function has two arguments, the first being the string `'10>15 is '`, which is printed *as is*, the second being the expression `10>15`, without quotes, which is evaluated to *False* before being printed.

The logical operators are

`and`	True if both the operands are true
`or`	True if either of the operands is true
`not`	True if operand is false (complements the operand)

for instance

```
>>> x=10
>>> y=15
>>> z=20
>>> print('x==y is ',x==y)
x==y is  False
>>> print('not x==y is ',not x==y)
not x==y is  True
>>> print(x!=y and y!=z)
True
>>> print(x<y or y>z)
True
```

1.8 Python Packages and the `import` Statement

We have already met a few Python built-in functions, like `print()`, `int()`, `float()`, ..., that we can use as soon as we enter Python's interactive mode. Python comes with a huge number of such predefined function which, apart from saving us the time of writing the functions ourselves (see Sect. 2.2), have the further, important advantage that they are optimized and stored in machine language, so that they are executed much faster than user-defined code. This is one of the strengths of Python. The number of Python predefined functions is so large that it would not be convenient to have all of them always automatically accessible. The functions are thus grouped into separate *modules*, which, in turn, may be grouped into *packages*, and one must import them from their respective packages before use. This is done with the `import` statement. The packages from which we shall import functions more often are called `math` (mathematical functions), `numpy` (the *numerical* package), `scipy` (the *scientific* package), and `matplotlib` (the plotting package). Packages `scipy` and `matplotlib` are further divided into subpackages. As an example, suppose that you need the square root of 5. If, after entering the interactive mode, you simply type `sqrt(5)` you get an error message

```
>>> sqrt(5)
Traceback (most recent call last):
  File "<stdin>", line 1, in <module>
NameError: name 'sqrt' is not defined
```

because the function `sqrt()` does not belong to the set of automatically available Python functions: you must import it from `math` (or from `numpy`) before use. For this you have the following three possibilities:

```
>>> import math as mt
>>> mt.sqrt(5)
2.23606797749979
```

here we have imported the whole `math` package under the name `mt`, and from now on we have access to all `math` functions by preceding their names with the prefix "`mt.`". Obviously you can replace the name `mt` with any name of your choice, provided you use it consistently in your following commands. After importing you can access any other `math` function, for instance you can type

```
>>> mt.cos(mt.pi)
-1.0
```

and get the cosine of π, `mt.pi` being the value of π stored in the `math` package (usually 3.141592653589793). Another import possibility is

```
>>> from math import sqrt
>>> sqrt(5)
2.23606797749979
>>> cos(pi)
Traceback (most recent call last):
  File "<stdin>", line 1, in <module>
NameError: name 'cos' is not defined
>>>
```

here we have imported *only* the function `sqrt()` from `math`, and we can used it without prefix. All other `math` functions have not been imported, therefore, for instance, the function `cos()` is not defined. You can use this method to import as many functions as you wish simultaneously, for instance

```
>>> from math import sqrt,cos,sin
>>> print(sqrt(5),cos(0),sin(0.3))
2.2360679775 1.0 0.295520206661
```

Finally we can type

```
>>> from math import *
>>> sqrt(5)
2.2360679774997898
>>> cos(pi)
-1.0
```

where the asterisk (`*`) stands for "everything". This command imports everything from `math`, and gives us access to all `math` functions without prefixes. However, this method is discouraged because it can lead to name collisions if used for more packages in the same session, or if a predefined variable of the imported package has the same name as one of your variables. The functions defined in `math` are listed in Appendix A, as well as the packages of the cmath module, comprising the definitions of the complex functions.

Other packages, like `numpy` and `scipy`, comprise huge numbers of functions and definitions each, and we cannot list them all in the present book. You can easily find the complete lists on the internet. Remember that often the same function is defined in different packages.

1.9 Conditional Statements

The comparison and logical operators usually appear in *conditional statements*. Conditional statements are vital in any programming language: they are needed whenever, starting from a given point of the program, we must follow different algorithms depending on whether, at that point, a condition evaluates to *True* or *False*. Conditional statements are written through the `if`, `elif` and `else` statements, which operate as you would expect from their names, `elif` standing for *else if*. This is a simple example

```
>>> from math import sqrt
>>> x=-5
>>> if x>0:
...     print("x is positive")
...     print(sqrt(x))
... elif x<0:
...     print("x is negative")
...     print(sqrt(-x))
... else:
...     print("x is zero")
...     print(0)
...
x is negative
2.2360679775
>>>
```

Note that all conditional statements end with a colon "`:`", and that the lines to be executed if the condition is true (the *conditioned* commands) are indented by the same amount of space with respect to the conditional statement. You can use spaces or tabs for indentation, but you are advised not to mix them: use only tabs, or only spaces, according to your taste. Now to the code. At the first prompt we import the function `sqrt()` from the `math` package. At the second prompt we create a variable `x` to which we assign a negative value. Starting from the third prompt we build an `if-elif-else` sequence that determines what to do if `x` is greater than, smaller than, or equal to 0. If `x` is greater than zero the two indented lines following the condition "`if x>0:`" are executed, and the rest of the code is ignored. If `x` is *not* greater than zero, the `elif` (*else if*) condition is checked, and if the condition is met, the `elif` conditioned code (the two following lines, indented relative to the statement "`elif x<0:`") are executed, and the following code is skipped. If neither the `if` nor the `elif` conditions are met, the code conditioned by `else` is executed. In an `if, elif, else` sequence there may be only one `if` (and there *must* be one!), there may be any number of `elif` conditions (including zero), and only one, or zero final `else`.

1.10 Lists and Tuples

A *list* is a a set of values (items), which is written as a sequence of comma-separated items between square brackets. The items in a list need not be of the same type.

Creating a list can be done by typing different comma-separated values between square brackets. For example

```
>>> list1 = ['physics', 'chemistry', 1997, 2000]
>>> list2 = [1, 2, 3, 4, 5 ]
>>> list3 = ["a", "b", "c", "d"]
>>> print(list2,list3)
[1, 2, 3, 4, 5] ['a', 'b', 'c', 'd']
>>> print(list3[1])
b
>>> list2[3]=10
>>> print(list2)
{[1, 2, 3, 10, 5]}
>>> list2=['cat','platypus']
>>> print(list2[1])
platypus
>>>
```

Similar to string indices, list indices start at 0, and lists can be sliced, concatenated and so on. Single list elements are accessed by indexing, like `list2[3]` above.

A *tuple* is a sequence of immutable Python objects. Tuples are sequences, just like lists. The relevant difference between tuples and lists is that a tuple, and its elements, cannot be changed. Syntactically, a tuple is declared by typing its comma-separated items between parentheses (round brackets, which, however, are not mandatory in the declaration), whereas lists use square brackets (which are mandatory). Thus you can create a tuple simply by typing different comma-separated variables, within parentheses or not.

```
>>> tuple1=4,5,6
>>> tuple2=('meerkat','walrus','carpenter')
>>> print(tuple1)
(4, 5, 6)
>>> print(tuple2[0])
meerkat
>>> tuple2[2]='elephant'
Traceback (most recent call last):
  File "<stdin>", line 1, in <module>
TypeError: 'tuple' object does not support item assignment
>>>
```

As stated above, changing a tuple is not allowed. However, it is possible to convert lists to tuples, and tuples to lists, analogously to what can be done between different number types or strings. For instance

```
>>> animals=list(tuple2)
>>> animals[2]='elephant'
>>> print(animals)
['meerkat', 'walrus', 'elephant']
>>>
```

here the elements of `tuple2` are copied into the list `animals`, which, being a list, is not immutable. If, eventually, we type

```
1 >>> tuple2=tuple(animals)
2 >>> tuple2
3 ('meerkat', 'walrus', 'elephant')
4 >>>
```

we see that, apparently, we have changed the tuple! Actually, Line 1 has destroyed the original tuple and created a new one with the same name, comprising the elements of the list `animals`.

1.11 List Methods

Python *methods* are functions that belong to Python objects. In Chap. 8 we shall meet methods belonging to instances of a *class*, here we consider methods belonging to list instances. Python includes the following *list methods*, that can change list instances (lists)

`append()`	`index()`	`remove()`
`count()`	`insert()`	`reverse()`
`extend()`	`pop()`	`sort()`

All above methods do what you can expect from their names, here are examples of how they operate

```
>>> animals=["cat","dog","goose"]
>>> animals.append("meerkat")
>>> animals
['cat', 'dog', 'goose', 'meerkat']
>>> animals.count("goose")
1
>>> animals.insert(2,"duck")
>>> animals
['cat', 'dog', 'duck', 'goose', 'meerkat']
>>> animals.insert(0,"bear")
>>> animals
['bear', 'cat', 'dog', 'duck', 'goose', 'meerkat']
>>> animals.index('dog')
2
>>> animals.index("snail")
Traceback (most recent call last):
  File "<stdin>", line 1, in <module>
ValueError: 'snail' is not in list
>>> popped=animals.pop()
>>> popped
'meerkat'
>>> animals
['bear', 'cat', 'dog', 'duck', 'goose']
>>>animals.reverse()
```

```
>>> animals
['goose', 'duck', 'dog', 'cat', 'bear']
>>> animals2=["meerkat","elephant","penguin"]
>>> animals.extend(animals2)
>>> animals
['goose', 'duck', 'dog', 'cat', 'bear', 'meerkat', 'elephant',\
'penguin']
>>> animals.sort()
>>> animals
['bear', 'cat', 'dog', 'duck', 'elephant', 'goose', 'meerkat',\
'penguin']
>>>
```

Method `append()` appends a new element at the end of the list, `count()` returns the number of times the argument occurs in the list, `extend()` appends another list at the end of the list, `index(obj)` returns the lowest index of `obj` in the list (remember that indices start from 0), `insert(index,obj)` inserts `obj` at position `index`, `pop()` returns the last item of the list and erases it from the list, `remove(obj)` removes `obj` from the list, `reverse()` reverses the list order and `sort()` sorts the list elements.

1.12 Lists and the = Assignment Operator

1.12.1 Copying Lists

Some care must be taken when using the = assignment operator with lists:

```
>>> a=[3,4,5,6]
>>> print(a)
[3, 4, 5, 6]
>>> b=a
>>> print(b)
[3, 4, 5, 6]
>>> a[0]=127
print('a=',a,'b=',b)
a= [127, 4, 5, 6] b= [127, 4, 5, 6]
>>>
```

The first statement creates a list comprising the numbers 3, 4, 5 and 6 as elements, and the variable `a` points to it. The statement `b=a` makes `b` point to exactly the same memory location as `a`. Thus, when the first element of the list is changed by the statement `a[0]=127`, the change affects both `a` and `b`. If you want to handle two independent lists, you must proceed this way

```
>>> a=[3,4,5,6]
>>> b=a.copy()
>>> a[0]=127
>>> print('a=',a,'b=',b)
a= [127, 4, 5, 6] b= [3, 4, 5, 6]
```

The method `.copy()` makes `b` point to an independent list, initially identical to the list pointed to by `a`, but any subsequent change to `a` does not affect `b`, and vice versa. Another possibility is

```
a=[3,4,5,6]
>>> b=a[:]
>>> a[0]=10
>>> print('a=',a,'b=',b)
a= [10, 4, 5, 6] b= [3, 4, 5, 6]
>>>
```

When in doubt, you can use the `id()` function, which returns the "identity" of its argument. An object identity is an integer unique for the given object, which remains constant during the object lifetime

```
>>> a=[3,4,5,6]
>>> b=a
>>> c=a.copy()
>>> print(id(a),id(b),id(c))
140402853321544 140402853321544 140402912014856
>>>
```

As stated above, the statement `b=a` makes `b` point to the same memory location as `a`, thus `b` and `a` have the same *identity*, while `c` points to an independent copy of `a`, thus its *identity* is different.

1.12.2 Copying the Elements of Lists and Tuples

It is possible to use the = assignment operator to copy the values of the elements of lists and tuples to separate variables in a single command. For instance, if you type

```
>>> a=[10,20,30]
>>> x1,x2,x3=a
>>> print(x2)
20
>>>
```

the variables `x1`, `x2` and `x3` are assigned the values of the three elements of the list `a`. The number of comma-separated values on the left-hand side **must** equal the number of elements of the list, or tuple, on the right-hand side, otherwise Python will report an error

```
>>> x4,x5=a
Traceback (most recent call last):
  File "<stdin>", line 1, in <module>
ValueError: too many values to unpack (expected 2)
>>>
```

As we know from Sect. 1.10, the use of parentheses is optional when assigning a tuple. Thus in the following code

```
>>> x1,x2,x3=15,'walrus',25
>>> print(x1,x2,x3)
15 walrus 25
%>>>
```

the first command can be considered both as a tuple unpacking or as a multiple assignment.

A list can comprise any number of elements, including 0 (empty list) and 1 (single-element list). Some care must be taken when unpacking a single-element list:

```
1 >>> a=[15]
2 >>> b=a
3 >>> c,=a
4 >>> print(b)
5 [15]
6 >>> print(c)
7 15
8 >>>
```

The first command creates the list a, comprising the single element 15. The command b=a does not copy the single element of a to a single variable b, rather, it makes b point to the same memory location as a, as we saw in Sect. 1.12.1. In order to copy the only element of a into a variable c we need the command at Line 3, c,=a. The comma after c tells Python that we are extracting the elements of the list a, the absence of variables after the comma tells that the list comprises a single element.

1.13 Slicing Lists and Strings

When you have a list, a tuple or an array it is possible to extract or modify specific sets of sub-elements without recurring to the loops that we shall encounter in Sect. 1.16. For instance, consider the list

```
a=[0,1,2,3,4,5,6,7,8,9,10,11,12]
```

and assume that you need a list b comprising the first 6 elements of a. This is how to do it in Python

```
>>> b=a[:6]
>>> b
[0, 1, 2, 3, 4, 5]
>>>
```

this command generates a new list b comprising a *slice* of a containing its first 6 elements. You can also type

```
>>> c=a[3:8]
>>> c
[3, 4, 5, 6, 7]
```

```
>>> d=a[4:]
>>> d
[4, 5, 6, 7, 8, 9, 10, 11, 12]
>>>
```

copying into c the *slice* of a from element number 3 (remember that the first element of the list is element number 0) up to, but not including, element number 8. For d we obtain the slice of the elements from the fifth (labeled by number 4) to the end. The lower and upper limits of the slice are separated by a colon inside the square brackets. The default values are 0 for the lower limit, and the whole list for the upper limit. Thus, b=a[:] copies the whole list a into b, as seen at the end of Sect. 1.12. Slicing can use a third argument (separated by a second colon), corresponding to a *step*. For instance

```
>>> a[::3]
[0, 3, 6, 9, 12]
>>>
```

here we have simply printed out the result, without storing it in a new variable. What we have got is a sublist of every third element of the list, starting from element 0 up to the last element. We can also use all the three arguments, for instance if we type

```
>>> a[1:11:2]
[1, 3, 5, 7, 9]
>>>
```

we select every second element (the third slicing argument is 2) starting from element number 1 (the second element of the list, specified by the first slicing argument), up to, but not including element number 11, specified by the second slicing argument.

Slicing of strings is perfectly analogous to slicing of lists:

```
>>> str='once upon a time'
>>> str[5:9]
'upon'
>>> str[::2]
'oc pnatm'
>>> str[::-1]
'emit a nopu ecno'
>>>
```

In the last case a negative step means going backwards, considering the string (or the list) as extended cyclically, and what we get is the string in reverse order.

Slicing can also be used to modify selected elements of lists. For instance, if we type

```
>>> thislist=[3,6,9,12,15,18,21]
>>> thislist[::3]=[24,96,168]
>>> thislist
[24, 6, 9, 96, 15, 18, 168]
>>>
```

every third element of thislist has been replaced by an element of the list [24,96,68]. It is important that the number of elements selected by slicing the list on the left hand side must exactly match the number of elements of the list at the right, otherwise Python reports an error

```
>>> thislist[::3]=[3,12,21,85]
Traceback (most recent call last):
  File "<stdin>", line 1, in <module>
ValueError: attempt to assign sequence of size 4 to extended slice\
of size 3
>>>
```

1.14 Dictionaries

A Python dictionary is an unordered collection of items. Each dictionary item is a pair consisting of a *key* and a *value*, with the requirement that the keys must be unique within one dictionary. The key and value of a dictionary item are separated by a colon (:), different items (different pairs) are separated by commas. A dictionary is created by writing its items within braces

```
>>> MyDict={'gatto':'cat','cane':'dog','ornitorinco':'platypus'}
>>> print(MyDict['cane'])
dog
>>>
```

a value of the dictionary `MyDict` can be retrieved by indexing `MyDict` with the corresponding key in square brackets. An alternative way of creating a dictionary is through the `dict()` function. For instance:

```
>>> MyOtherDict=dict(Katze='cat',Hund='dog',Schnabeltier='platypus')
>>> MyOtherDict
{'Katze': 'cat', 'Hund': 'dog', 'Schnabeltier': 'platypus'}
>>> print(MyOtherDict['Katze'])
'cat'
>>>
```

In this case the arguments of the `dict()` function are written within parentheses, as usual for functions. Keys are not written within quotes, each *key* is separated by the corresponding *value* by an equal sign (=) rather than by a colon (:).

A new dictionary item is added by simply assigning it

```
>>> MyDict['ape']='bee'
>>> print(MyDict)
{'ape': 'bee', 'ornitorinco': 'platypus', 'cane': 'dog',\
'gatto': 'cat'}
>>>
```

A dictionary item can be removed with the `del` statement

```
>>> del MyDict['ornitorinco']
>>> print(MyDict)
{'ape': 'bee', 'cane': 'dog', 'gatto': 'cat'}
>>>
```

you can modify an existing item by reassigning it

```
>>> MyDict['cane']='hound'
>>> print(MyDict)
{'ape': 'bee', 'cane': 'hound', 'gatto': 'cat'}
>>>
```

Dictionary values have no restrictions: a dictionary item can be any arbitrary Python object, either a standard object or a user-defined object. However, the same is not true for the keys: duplicate keys are not allowed, and keys are immutable. This means that you can use strings, numbers or tuples as dictionary keys, but not, for instance, lists.

1.15 The `print()` Function and Formatting

We have already met the `print()` function with single and multiple arguments of different types (integer and float numbers, strings, lists, ...) in the previous sections of this chapter. Its purpose is to print its arguments on the terminal. Note that in interactive mode the value of any variable can also be printed by simply typing its name at the Python prompt, as we have seen in many previous examples. Here follows an example of the use of the `print()` function

```
>>> animal='platypus'
>>> num=4.5
>>> print(animal,num,27,4/3)
platypus 4.5 27 1.3333333333333333
>>>
```

Often we need a more refined way of printing, which, for instance, allows us to control the number of digits printed after the decimal point for a float number like 4/3. Or we might want to align the decimal points of numbers printed in successive lines. All this, and more, is achieved through *formatted printing*. For this Python offers two possibilities.

1.15.1 Old Style

The old style format is very similar to the C/C++ format, for instance:

```
>>> from math import sqrt
>>> num=15/7
>>> print('the square root of %.5f is %.5f'%(num, sqrt(num)))
the square root of 2.14286 is 1.46385
>>>
```

The string `'the square root of %.5f is %.5f'` appearing as argument of the `print()` function at the third line is a *format string*. A format string, written within quotes like all strings, is printed by the `print()` function on the terminal as is, character by character, except for the *placeholders* it may contain. Placeholders are substrings beginning with the `%` character, here we have two of them, both in the form `%.5f`. The format string is followed by a *string modulo operator*, represented by a `%` character, which couples the format string to a following tuple. The tuple, `(num,sqrt(num))` in our case, comprises the values to be inserted at the locations of the placeholders. Thus, the number of elements of the tuple must equal the number of placeholders in the format string. The general syntax for a placeholder is

`%[flags][width][.precision]type`

The parts within square brackets are optional, while the leading `%` and the type are mandatory. In the case of `%.5f` neither *flags* nor *width* are given, while the *precision*, `.5`, requires 5 digits after the decimal point. The *type* `f` indicates that the tuple elements to be printed are float numbers. The following lines show how formatted printing can be used for aligning numbers of successive lines

```
>>> print("%5d %10.5 f\n %5d %10.5f" %(11, sqrt(11), 1525, sqrt(1525)))
   11    3.31662
 1525   39.05125
>>>
```

here the placeholder `%5d` requires 5 characters of width for an integer number (*type* `d`), while the placeholder `%10.5f` requires a *width* of 10 characters (characters include the decimal point, and, for a negative number, the minus sign), with 5 digits after the decimal point (*precision*) for a float number (type `f`). When the width is specified, the numbers are right-justified within the reserved space (10 characters in the case of `%10.5f`), with trailing blanks added at the left in order to complete the width. The newline code `\n` in the format inserts a *new line* command, thus splitting the output into two consecutive lines.

1.15.2 New Style

Also the new-style format makes use of a format string as argument of the `print()` function. But the format string is not coupled to a tuple of variables via the string modulo operator. Rather, the `.format()` method (we recall that a Python *method* is a function belonging to the object preceding the dot) is applied to the format string. The format string still has placeholders for the variables to be inserted. The syntax of the placeholders is similar to the old-style syntax, but the percent character, `%`, is replaced by a colon character, `:`, and placeholders are written within braces. This is an example

```
>>> x=1525
>>> print("the square root of {:5d} is {:10.5f}".format(x,sqrt(x)))
the square root of  1525 is   39.05125
>>>
```

The new-style allows us to use optional positional parameters before the colons in the placeholders, so that the format arguments can be written in any order

```
>>> x=1525
>>> print("the square root of {a:5d} is {b:10.5f}".format(b=sqrt(x),a=x))
the square root of 1525 is    39.05125
```

where a and b are positional parameters. Any number of positional parameters is allowed. Positional parameters also make multiple use of a single variable possible:

```
>>> print("{k:.2f}, or, more precisely, {k:.10f}".format(k=17/13))
1.31, or, more precisely, 1.3076923077
```

1.15.3 Format Types and Flags

These are the meanings of the conversion types appearing in the format placeholders

d	signed integer decimal	f	floating point decimal format
i	signed integer decimal	F	floating point decimal format
o	unsigned octal	g	same as e or f, see below
x	unsigned hexadecimal (lower case)	G	same as E or F, see below
X	unsigned hexadecimal (upper case)	c	single character
e	floating point exponential (lower case)	r	string
E	floating point exponential (upper case)	s	string

Types d, i, o, x and X refer to integer numbers. Types d and i print signed integers in decimal format, and are fully equivalent. Type o prints unsigned integers in octal format, while types x and X print unsigned numbers in hexadecimal format, x prints the hexadecimal digits A-F in lower case, X prints the same hexadecimal digits in upper case. This is an example

```
>>> print("{a:d}  {a:o}  {a:x}  {a:X}".format(a=254))
254  376  fe  FE
>>>
```

Note that $3 \times 8^2 + 7 \times 8 + 6 = 254$, and $15 \times 16 + 14 = 254$. Types e, E, f, F, g and G are for float numbers. A placeholder {:w.pe}, or {:w.pE}, w being the required *width* and p the required *precision*, causes the corresponding number to be printed in *scientific notation* , i.e., in the form $x \times 10^n$, where n is an integer, and x is a real number such that $1 \leqslant |x| < 10$. A placeholder {:w.pf} prints the number in normal float representation. For instance

```
>>> print("{b:10.4f} {b:10.4e} {b:10.4E}".format(b=1527.42))
 1527.4200 1.5274e+03 1.5274E+03
>>>
```

The width includes the characters needed for the decimal point, the minus sign if the number is negative, and, in the case of scientific notation, also the exponent. The only difference between e and E types is the case of the letter e (E). There is no difference between f and F. Type g (G) is equivalent to type e (E) if the exponent is greater than -4 or less than p, equivalent to f (F) otherwise. See the following two examples

```
>>> print("{a:15.5f}{a:15.5e}{a:15.5g}".format(a=5000/3))
     1666.66667    1.66667e+03         1666.7
>>> print("{a:15.5f}\{a:15.5e}{a:15.5g}".format(a=5e-5/3))
        0.00002    1.66667e-05     1.6667e-05
>>>
```

Type `c` prints a single character, and accepts an integer (ASCII or UTF-8 character encoding) as format argument

```
>>> for i in range(65,70):print("{:c}".format(i))
...
A
B
C
D
E
>>>
```

Here we have used a `for` loop, to be discussed in Sect. 1.16. Types `r` and `s` refer to strings.

```
>>> print("{:s}".format("Have a good day!"))
Have a good day!
>>> print("{!r}".format("Have a good day!"))
'Have a good day!'
>>>
```

Note that the string is printed within quotes if you use the `r` code, and that the `r` code must be preceded by an exclamation mark, `!`, rather than by a colon, `:`.

1.16 Loops

Loops are extremely important in computer programming. Python has two types of loops: the `for` and the `while` loop.

1.16.1 The `for` Loop

It is important to realize that the Python `for` loop behaves differently from the `for` loop of other programming languages, notably C/C++. For using the Python `for` loop you must first have a sequence of elements, for instance a list or a tuple, and the `for` loop iterates over the elements of the sequence. This is an example

```
>>> flowers=["rose","cyclamen","daisy","tulip"]
>>> for x in flowers:
...     print(x)
...
rose
cyclamen
```

```
daisy
tulip
>>> from math import sqrt
>>> numbers=[533,712,925]
>>> for x in numbers:
...      print("{:10.4f}".format(sqrt(x)))
...      print("this was the square root of {:10.4f}".format(x))\
...
   23.0868
this was the square root of   533.0000
   26.6833
this was the square root of   712.0000
   30.4138
this was the square root of   925.0000
>>>
```

The declaration of a `for` loop must terminate with a colon (`:`), and all commands of the loop must be indented, for instance by a `tab`, or by an arbitrary, but fixed number of spaces, relatively to the line of the loop declaration. In interactive mode, the command sequence of a `for` loop (or of a `while` loop) is terminated by typing an unindented empty line. The iteration sequence of a `for` loop can be provided by the `range()` function, which accepts up to three arguments

```
>>> for x in range(5):
...      print(x)
...
0
1
2
3
4
>>> for x in range(6,10):
...      print(x)
...
6
7
8
9
>>> for x in range(6,12,3):
...      print(x)
...
6
9
>>>
```

Note that the `range()` function is zero-based, thus, `range(5)` generates the list of the first 5 integer numbers [0, 1, ... 4], while the expression `range(`n_1`,`n_2`)` generates the list of the $n_2 - n_1$ integers $[n_1, n_1 + 1, \ldots, n_2 - 1]$. Finally, in the case of three arguments, `range(`n_1`,`n_2`,`n_3`)`, the third argument is interpreted as the constant spacing between successive numbers of the list. The last number of the list is the highest number smaller than n_2. This behavior is anologous to the behavior of slicing that we met in Sect. 1.13.

It is also important to remember that the `range()` function is *not* part of the `for` syntax. Rather, `range()` is a built-in Python function that creates a number list. The number list is created *before* the loop is executed, and it cannot be changed during the loop execution. Thus, for instance, if you type

```
>>> a=3
>>> for i in range(a):
...         print(i)
...         if i==0:
...         a=10
...
0
1
2
>>> print(a)
10
>>>
```

the loop iterates over the list `[0,1,2]` in spite of the fact that the value of `a` is changed at the first iteration. However, the `for` loop can be interrupted by a `break` statement, as discussed in Sect. 1.16.3.

Another important built-in Python function, particularly useful when associated to a `for` loop, is `enumerate()`. It allows us to loop over a sequence and have a simultaneous automatic counter. Here is an example of its use

```
>>> flowers=[rose,cyclamen,daisy,tulip]
>>> for i,x in enumerate(flowers):
...     print(i,x)
...
0 rose
1 cyclamen
2 daisy
3 tulip
>>>
```

It is also possible to have the counter starting from any given integer value, by adding an optional argument to `enumerate()`

```
>>> for i,x in enumerate(flowers,15):
...     print(i,x)
...
15 rose
16 cyclamen
17 daisy
18 tulip
>>>
```

Obviously, variables `i` and `x` are available for any possible use in the loop commands.

1.16.2 The `while` Loop

A `while` loop iterates as long as a specified Boolean condition is true. For example:

```
>>> i=5
>>> while i<=8:
...     print(i)
...     i+=1
...
5
6
7
8
>>>
```

Here the variable `i` is set equal to 5 before the loop begins. The loop is iterated till i is smaller than, or equal to 8, each iteration prints the value of `i`, then increments it by 1.

1.16.3 Breaking a Loop

A loop can be interrupted by a `break` statement. For instance, we might write for a `for` loop

```
>>> for j in range(10):
...     if j>3:
...         break
...     print(j)
...
0
1
2
3
>>>
```

The `break` is executed when the condition `j>3` is met. Note the double indentation before the `break` command: an indentation relative to the loop, plus an indentation relative to the `if` statement. The `print(j)` line has a single indentation, since it is part of the loop, but it is not conditioned by the `if` statement.

The `break` statement has an interesting application in the `while` loop:

```
>>> count=0
>>> while True:
...     print(count)
...     count+=1
...     if count>2:
...         break
...
0
1
2
>>>
```

Here the Boolean condition for the `while` loop is always true by definition, being the Boolean constant *True* itself. Thus, the loop would go on forever if it did not contain the `break` condition.

1.16.4 The `continue` Statement

The `continue` statement is used to skip the part of the loop (either `for` or `while`) that follows it, without interrupting the loop itself, but passing to the following iteration. For instance, the following code skips all odd numbers, printing only the even ones

```
>>> for i in range(5):
...     if i%2>0:
...         continue
...     print(i)
...
0
2
4
```

If the condition `i%2>0` is met, i.e., if the remainder of the division of `i` by 2 is greater than zero (`i` is odd), the print function is skipped, and the next iteration is started.

1.17 Operations with Matrices and Vectors

1.17.1 Lists and Arrays

In a sense, a vector is a list of numbers, and a matrix is a list of lists of numbers, so, we can write

```
>>> v1=[1,2,3]
>>> v2=[4,5,6]
>>> a1=[[1,2,3],[4,5,6],[7,8,9]]
>>> a2=[[9,8,7],[6,5,4],[3,2,1]]
>>> print(v1)
[1, 2, 3]
>>> print(a1)
[[1, 2, 3], [4, 5, 6], [7, 8, 9]]
```

However, we get a surprise if we add two *such* vectors, or two *such* matrices,

```
>>> print(v1+v2)
[1, 2, 3, 4, 5, 6]
>>> print(a1+a2)
[[1, 2, 3], [4, 5, 6], [7, 8, 9], [9, 8, 7], [6, 5, 4], [3, 2, 1]]
```

the surprise being due to the fact that, when two Python *lists* are added, the result is a new list comprising the elements of both lists. If we want the usual mathematical operations on matrices and vectors, we need `numpy.array` objects instead of lists. Thus, first we must import `numpy`, then declare vectors and matrices as `numpy.arrays`

```
>>> import numpy as np
>>> v1=np.array([1,2,3])
>>> v2=np.array([4,5,6])
>>> a1=np.array([[1,2,3],[4,5,6],[7,8,9]])
>>> a2=np.array([[9,8,7],[6,5,4],[3,3,1]])
>>> print(a1)
[[1 2 3]
 [4 5 6]
 [7 8 9]]
>>> a1
array([[1, 2, 3],
       [4, 5, 6],
       [7, 8, 9]])
>>> print(v1+v2)
[5 7 9]
>>> print(a1+a2)
[[10 10 10]
 [10 10 10]
 [10 11 10]]
>>> print(a1+v1)
[[2 4 6]
 [5 7 9]
 [8 10 12]]
```

Note the output difference after typing `print(a1)` and simply typing `a1`: in the latter case, the word `array()` appears, and commas are printed between row elements. Adding two arrays leads to the expected result if the two arrays have the same dimensions. However it is possible to add a matrix and a vector, an operation not allowed in usual matrix algebra, provided that the number of columns of the matrix equals the number of elements of the vector. The vector elements are added to the corresponding elements of each matrix row.

This is a first step into matrix algebra, but we still get a surprise if we multiply two arrays by using the usual `*` multiplication operator, met in Sect. 1.5

```
>>> print(v1*v2)
[ 4 10 18]
>>> print(a1*a2)
[[ 9 16 21]
 [24 25 24]
 [21 24  9]]
>>> print(a1*v1)
[[ 1  4  9]
 [ 4 10 18]
 [ 7 16 27]]
```

what we get is actually an *element by element multiplication* (sometimes called *Hadamard product*), as shown above. Note that the `*` multiplication of a matrix by a vector multiplies the elements of each row of the matrix by the corresponding elements of the vector. It is also possible to obtain a Hadamard element by element division, using either the / operator or the `numpy.divide()` function

```
>>> np.divide(v1,v2)
array([ 0.25,  0.4 ,  0.5 ])
>>> print(v1/v2)
[ 0.25  0.4   0.5]
```

Note that the `numpy` package contains also many mathematical functions, sharing the names with the `math` functions. If you apply these `numpy` functions to *scalar* quantities, their behavior is analogous to the behavior of their `math` counterparts, with a possible difference in precision

```
>>> import numpy as np
>>> import math as mt
>>> print(mt.sqrt(5),np.sqrt(5))
2.23606797749979 2.2360679775
>>> print(mt.sin(mt.pi/4),np.sin(np.pi/4))
0.7071067811865475 0.707106781187
```

But there is an important difference: the `numpy` mathematical functions can take *arrays*, or lists, as arguments , while the `math` functions can't,

```
>>> a=[5,6,7,8]
>>> np.sqrt(a)
array([ 2.23606798,  2.44948974,  2.64575131,  2.82842712])
>>> mt.sqrt(a)
Traceback (most recent call last):
  File "<stdin>", line 1, in <module>
TypeError: must be real number, not list
```

We shall see the importance of the `numpy` mathematical functions in Chap. 3, where we shall discuss how to plot functions. As an example, consider the following commands

```
>>> import numpy as np
>>> x=np.arange(0,3.6,0.5)
>>> x
array([ 0. ,  0.5,  1. ,  1.5,  2. ,  2.5,  3., 3.5 ])
>>> y=np.sin(x)
>>> y
array([ 0.        ,  0.47942554,  0.84147098,  0.99749499,  0.90929743,\
        0.59847214,  0.14112001, -0.35078323])
```

Command `x=np.arange(0,3.6,0.5)` generates an array `x` of floats ranging from 0.0 to 3.5, which we could use as abscissae of a plot. Command `y=np.sin(x)` generates an array `y` such that each of its elements is the sine of the corresponding element of `x`. Thus, the elements of `y` can be used as ordinates of our plot.

1.17.2 Slicing out Rows and Columns from a Matrix

Slicing out a row from a matrix is obvious:

```
>>> a1=np.array([[1,2,3],[4,5,6],[7,8,9]])
>>> a1
array([[1, 2, 3],
       [4, 5, 6],
       [7, 8, 9]])
>>> row0=a1[0]
>>> row0
array([1, 2, 3])
>>> row1=a1[1]
>>> row1
array([4, 5, 6])
```

Slicing out a column is slightly less obvious:

```
>>> column0=a1[:,0]
>>> column0
array([1, 4, 7])
>>> column2=a1[:,2]
>>> column2
array([3, 6, 9])
```

note the comma before the index.

1.17.3 Arrays and Matrix Arithmetics

If what we want is the usual row-by-column matrix multiplication we must use the `numpy.dot()` function, which performs the scalar (or *dot*) product of two arrays. A few examples follow

```
>>> c=np.dot(v1,v2)
>>> print(c)
32
>>> a3=np.dot(a1,a2)
>>> print(a3)
[[ 30  27  18]
 [ 84  75  54]
 [138 123  90]]
>>> v3=np.dot(a1,v1)
>>> print(v3)
[14 32 50]
>>> v4=np.dot(v1,a1)
>>> print(v4)
[30 36 42]
```

Once we have an array `a` we can use the function `numpy.tile()` to obtain a new array by repeating `a` a given number of times

```
>>> from numpy import tile
>>> a=[1,2,3,4]
>>> b=tile(a,3)
>>> b
array([ 1,  2,  3,  4,  1,  2,  3,  4,  1,  2,  3,  4])
>>> c=[[1,2],[3,4]]
>>> d=tile(c,3)
>>> d
array([[1, 2, 1, 2, 1, 2],
       [3, 4, 3, 4, 3, 4]])
```

the second argument of `tile()` giving the number of repetitions. Note that we built `a` and `c` as *lists*, but `b` and `d` are *arrays*. We can obtain a matrix from a vector, or, more generally, a final array of higher dimension than the start array, by using a tuple as second argument of `tile()`:

```
>>> c=tile(a,(len(a),1))
>>> c
array([[ 1,  2,  3,  4],
       [ 1,  2,  3,  4],
       [ 1,  2,  3,  4],
       [ 1,  2,  3,  4]])
```

the first item of the tuple `(len(a),1)` telling how many rows the matrix will have (here we are generating a square matrix), and the second how many times the array `a` is repeated in each row, here only once.

1.17.4 Further Matrix Operations

1.17.4.1 Transpose

We can transpose a matrix by using the `.transpose()` method:

```
d=c.transpose()
>>> d
array([[ 1,  1,  1,  1],
       [ 2,  2,  2,  2],
       [ 3,  3,  3,  3],
       [ 4,  4,  4,  4]])
```

The same result can be obtained by typing

```
>>> d=c.T
```

the `.T` operator being a shorthand for the `.transpose()` operator. Finally, we can generate a symmetric matrix `e` by typing, for instance

```
>>> e=d*a
>>> e
array([[ 1,  2,  3,  4],
```

```
       [ 2,  4,  6,  8],
       [ 3,  6,  9, 12],
       [ 4,  8, 12, 16]])
```

where $e_{ij} = a_i a_j$, since the `*` product multiplies each element of each row of `d` by the corresponding element of `a`, as discussed in Sect. 1.17.1. Analogously, we can generate a skew-symmetric matrix `f` by typing, for instance

```
>>> f=d-a
>>> f
array([[ 0, -1, -2, -3],
       [ 1,  0, -1, -2],
       [ 2,  1,  0, -1],
       [ 3,  2,  1,  0]])
```

where $f_{ij} = a_i - a_j$, since the `-` operator subtracts the elements of `a` from the corresponding elements of each row of `d`. Obviously, all the above operations can be obtained also by using nested `for` or `while` loops. For instance, our skew-symmetric matrix can be generated by the nested loops

```
>>> d=np.empty([len(a),len(a)])
>>> for i in range(len(a)):
...     for j in range(len(a)):
...         d[i,j]=a[i]-a[j]
...
```

where the `numpy.empty()` function returns a new array of given dimensions, without initializing the entries. Note that the function `numpy.empty()` has a single argument, which must be a tuple if a multidimensional array is needed. The matrix elements are then initialized by the nested loops. A more efficient way of writing the nested loops could be

```
>>> d=np.zeros([len(a),len(a)])
>>> i=0
>>> while i<len(a):
...     j=0
...     while j<i:
...         d[i,j]=a[i]-a[j]
...         d[j,i]=-d[i,j]
...         j+=1
...     i+=1
...
```

where the function `numpy.zeros()` returns a new array of given dimensions whose elements are all zero. As for function `numpy.empty()`, the single argument of `numpy.zeros()` must be a tuple if what we want is a multidimensional array. The skew symmetry of the matrix is exploited in order to skip the calculation of the diagonal elements and not to repeat the off-diagonal calculations twice. However, it is important to remember that Python array-manipulation functions are pre-compiled and optimized, thus they run much faster than the loops that a user can write in Python.

1.17.4.2 Sum the Elements over a Given Axis

Function `numpy.sum()` sums the elements of an array over a given axis.

```
>>> a=np.array([[1,2,3],[4,5,6],[7,8,9]])
>>> a
array([[1, 2, 3],
       [4, 5, 6],
       [7, 8, 9]])
>>> np.sum(a,axis=0)
array([12, 15, 18])
>>> np.sum(a,axis=1)
array([ 6, 15, 24])
```

axis 0 is the horizontal axis, parallel to the matrix rows, and axis 1 is the vertical axis, parallel to the matrix columns. Note that the argument `axis=0` in the `sum()` function indicates that columns are added, leaving a single, collapsed row. This can seem misleading when working with two-dimensional arrays (matrices), since `axis=0` means that sums are performed column-wise. However, this is the most straightforward definition when working with three- or higher-dimensional arrays.

1.17.4.3 Trace

The *trace* of a matrix is calculated by the `numpy.trace()` operator

```
>>> np.trace(a)
15
>>>
```

1.17.4.4 Determinant and Inverse

The *determinant* of a matrix is calculated by the `numpy.linalg.det()` operator, while the *inverse* of a matrix is evaluated by the `numpy.linalg.inv()` function

```
>>> a=np.array([[1,3,3],[3,2,1],[18,16,14]])
>>> a
array([[ 1,  3,  3],
       [ 3,  2,  1],
       [18, 16, 14]])
>>> np.linalg.det(a)
-23.999999999999993
>>> b=np.linalg.inv(a)
>>> b
array([[-0.5       , -0.25      ,  0.125     ],
       [ 1.        ,  1.66666667, -0.33333333],
       [-0.5       , -1.58333333,  0.29166667]])
>>> np.dot(a,b)
array([[  1.00000000e+00,   8.88178420e-16,   0.00000000e+00],
```

```
       [ 0.00000000e+00,  1.00000000e+00,  5.55111512e-17],
       [ 0.00000000e+00,  0.00000000e+00,  1.00000000e+00]])
>>>
```

Note the effects of rounding.

1.17.4.5 Diagonalization

The diagonalization of a matrix can be performed by the function `numpy.linalg.eig()`, which returns a tuple consisting of a vector and an array. The vector contains the eigenvalues in arbitrary order, while the array contains the corresponding eigenvectors, organized in a matrix, one eigenvector per column. The eigenvectors are normalized so their Euclidean norms are 1. An example of the use of `numpy.linalg.eig()` follows

```
1  >>> import numpy as np
2  >>> a=np.array([[1,2,3,4],[2,2,4,8],[3,4,5,6],[4,9,6,8]])
3  >>> a
4  array([[1, 2, 3, 4],
5         [2, 2, 4, 8],
6         [3, 4, 5, 6],
7         [4, 9, 6, 8]])
8  >>> eigval,eigvec=np.linalg.eig(a)
9  >>> eigval
10 array([19.53241835, -4.10596077, -0.56799313,  1.14153556])
11 >>> eigvec
12 array([[ 0.27701336,  0.18643609, -0.90273601, -0.26017229],
13        [ 0.45869181,  0.75119053,  0.15881403,  0.32137161],
14        [ 0.47236893,  0.02548963,  0.39903646, -0.80179608],
15        [ 0.69980927, -0.63269633, -0.0248134 ,  0.43145535]])
16 >>> np.dot(a,eigvec[:,0])
17 array([ 5.41074085,  8.95936022,  9.22650759, 13.66896747])
18 >>>
```

In the above example, Line 8 unpacks the output of `numpy.linalg.eig()`, storing the eigenvalues of the matrix `a` defined at Line 2 in the array `eigval`, and the eignevectors in the array `eigvec`. Line 16 multiplies matrix `a` by its first eigenvector (the first column of `eigvec`), sliced out of `eigvec`, see Sect. 1.13 on slicing. The result shown in Line 17 is the first eigenvector itself multiplied by the first eigenvalue (19.53241835). We can check this by typing the command

```
>>> np.dot(eigvec[:,0],eigval[0])
array([ 5.41074086,  8.95936024,  9.2265076 , 13.66896746])
```

which multiplies the first eigenvector by the first eigenvalue.

1.18 Exiting Interactive Mode

We exit from the Python interactive mode by typing either `quit()` (the parentheses are mandatory, in order to tell the interpreter that we are referring to a function), or Ctrl-D at the Python prompt.

1.19 Exercises

Exercise 1.1 Use interactive Python to solve the quadratic equation

$$2x^2 - 17x + 35 = 0 \,.$$

Exercise 1.2 Use interactive Python to count the digits in the product

$$99999 \times 78965321 \times 123456499 \times 987645 \,.$$

Exercise 1.3 Create a list containing the words *Aardvark, Aargau, Abacus, Blowfish, Caper,* then insert the word *Abalone* in alphabetical order.

Exercise 1.4 Write the matrix

$$\begin{pmatrix} 3.5 & 6 & 7 \\ 6 & 4.5 & 8 \\ 7 & 8 & 5.5 \end{pmatrix} \,.$$

then slice out the second column and the last row.

Exercise 1.5 Write a `while` loop that prints the first 10 integer numbers and their square roots in two columns. Align the decimal points.

Exercise 1.6 Write a `while` loop that prints the first 10 integers whose last hexadecimal digit is a letter in the range A–F.

Exercise 1.7 Multiply, row by column, the two matrices

$$A = \begin{pmatrix} 1 & 2 & 3 \\ 4 & 5 & 6 \\ 7 & 8 & 9 \end{pmatrix} \,, \qquad B = \begin{pmatrix} 4 & 6 & 5 \\ 9 & 8 & 7 \\ 7 & 5 & 3 \end{pmatrix} \,.$$

Exercise 1.8 In a Cartesian reference frame the vector $\boldsymbol{v} = (1, 2, 3)$ is rotated by 20° around the z axis. Evaluate its components after rotation.

Exercise 1.9 Build a multiplication table up to 9×9.

Exercise 1.10 Evaluate the eigenvalues and eigenvectors of the rotation matrix

$$R = \begin{pmatrix} \frac{\sqrt{2}}{2} & -\frac{\sqrt{2}}{2} & 0 \\ \frac{\sqrt{2}}{2} & \frac{\sqrt{2}}{2} & 0 \\ 0 & 0 & 1 \end{pmatrix} .$$

Chapter 2
Python Scripts

2.1 Scripts

A Python script is a text file, either ASCII or UTF-8, containing your Python code. A first possibility, if you choose to use the terminal, is typing the code of your script into a file with your favorite text editor (provided that you use it in plain text mode), and storing it in the computer memory. A second possibility, if you chose to download Anaconda, is typing the script in the Spyder left subwindow, see Fig. 1.2. In either case Python reads the file as a whole, and then executes its commands, rather than reading and executing the lines you type one by one, as it does in interactive mode. Scripts have some very important advantages over the interactive mode: for instance you do not have to retype a long code that you use in separate work sessions, and you can easily change a line of code without retyping all the rest. It is customary, and convenient, to give your script files names with a final "`.py`" extension, for instance `myscript.py`.

2.1.1 Running Scripts on a Terminal

Once you have typed and saved a script, you can run it in the terminal simply by typing

```
python3 myscript.py
```

The command `python3` followed by a file name does not enter the interactive mode, but reads the file, interprets and executes it. In Linux, an alternative is making your script file directly executable. For this, the first line of your file must be

---

Electronic supplementary material The online version of this chapter (https://doi.org/10.1007/978-3-030-45027-4_2) contains supplementary material, which is available to authorized users.

G. Moruzzi, *Essential Python for the Physicist*,
https://doi.org/10.1007/978-3-030-45027-4_2

```
#!/usr/bin/env python3
```

which states that the file must be handled by the python3 interpreter. The file, once written, is made executable by typing

```
chmod +x myscript.py
```

on the terminal, where `chmod` stands for "change mode" (change the *access permissions*) in Unix-like operating systems, and `+x` stand for "add the *execute* permission". Obviously, you must replace `mypscript.py` by the actual name of your script! Once the script is made executable, you run it simply by typing its name

```
myscript.py
```

An executable script is executed also if you type `python3 myprogram.py`. In fact python3, if called directly, interprets the first line, beginning with a hash sign (#), as a comment, and ignores it. If you are using Python 2 instead of Python 3, it is convenient to add a second "commented" line to the file

```
#coding: utf8
```

this tells the interpreter to use the more complete UTF-8 character encoding rather than plain ASCII characters. Python 3 uses UTF-8 encoding by default.

2.1.2 Running Scripts on Spyder3

If you typed your script in the editor subwindow (left subwindow of Fig. 1.2) of Spyder, you run it by first clicking on the `Run` option on the top menu bar. This will open a menu window, where you choose the first option: `Run`.

2.1.3 A First Script

As an example, let us write a script that evaluates the prime factors of an integer number. We can call this script `factorize.py`. As we already know, a hash symbol (#) followed by any character sequence is considered as a comment and ignored by Python. Thus, you can omit all the lines comprising only a hash symbol when you copy the file of Listing 2.1. We have inserted them only to separate code parts visually, in order to make the script more "human understandable". Further, if you copy this script, or any script listed in the rest of this book, you should *not* type the numbers at the beginning of each line: they are *not* part of the code. We have inserted them only to identify the lines in the code discussion.

Listing 2.1 factorize.py

```
#!/usr/bin/env python3
import math
import time
from sys import argv
#
```

We import the libraries *math* and *time*, and the list `argv` from the library *sys*. We shall see in the following why we need this stuff.

```
6 num = argv[1]
7 start = time.time()
8 #
```

Variable `argv` is a list of strings comprising everything that you typed in the command line when calling the script, including the script name itself. Thus, for instance, if you typed
`factorize.py 3512`
on the terminal, element `argv[0]` contains the string `'factorize.py'`, and `argv[1]` contains the string `'3512'`. The second string, `argv[1]`, is copied into the variable `num` at Line 6. If, on the other hand, you are using Spyder, once you have your code in the editor subwindow, you first click on the `Run` option on the top menu bar. This opens a menu window where you click on "`Command line options:`", then you type the number, for instance `3512`, in the window at the right of "`Command line options:`". Finally You click on the `Run` option at the bottom of the menu window.

At Line 7, the function `time.time()` returns the time in seconds elapsed since a platform-dependent starting instant, as a floating point number. The starting instant (*time origin*) for Unix and Unix-like systems is January 1st, 1970, at 00:00:00 (UTC). In any case, the actual time origin is not relevant: we are interested only in the *time difference* between the start and the end instants of the execution of our program, in order to know how long the program runs.

```
9 n=int(num)
10 print("digits:␣{:d}".format(len(num)))
11 #
```

At line 9 the string `num` (the UTF-8 string "3512" in our case) is converted into the integer variable `n`, which shall be used for computation. Line 10 counts the number of digits in the number to be factorized (4 in our case), and prints it on the terminal. Note that the symbol ␣ stands for a space (a blank).

```
12 sqnf=int(math.ceil(math.sqrt(float(n))))
13 #
14 factors = []
15 #
```

Line 12 stores in `sqnf` the maximum integer which can be a prime factor of `n`, i.e., the smallest integer greater than $\sqrt{n}$. Line 14 creates the empty list `factors`, that we shall fill with the prime factors of our number.

```
16 while n%2==0:
17     factors.append(2)
18     n=n//2
19     sqnf=int(math.ceil(math.sqrt(float(n))))
20 #
```

The `while` loop 16–18 stores 2 into `factors` if `n` is even. In this case `n` is replaced by its half, and Line 19 replaces `sqnf` by the value of the new largest possible remaining prime factor. The procedure is iterated as long as the resulting n is even, and each time a copy of the number 2 is stored into `factors`.

```
21  i=3
22  while i<=sqnf:
23     while n%i==0:
24        factors.append(i)
25        n=n//i
26        sqnf=int(math.ceil(math.sqrt(float(n))))
27     i+=2
28  #
```

Line 21 assigns the integer value 3 to `i`. Lines 22–27 are two `while` *nested loops*: the loop 23–26 is nested inside the loop 22–27. The loop of lines 23–26 is analogous to the loop 16–19, with 2 replaced by the successive odd numbers `i` generated at line 27, belonging to the external `while` loop. Of course, not all odd numbers are primes!!! But, if an odd number `i` is not prime, `n` has already been divided by the prime factors of `i`, and is no longer divisible by `i`. Thus, checking divisibility by all odd numbers is, in a sense, a more or less unavoidable waste of time, but does not lead to errors. Number `i` is stored into `factors` as long as `i` divides `n`.

```
29  if n!=1:
30     factors.append(n)
31  #
```

Now we are out of all factorizing loops. If the last value of `n` is different from 1, it must be a prime number, and must be stored into `factors`.

```
32  print(factors)
33  elapsed = time.time() - start
34  print("elapsed time: {:f} s\n".format(elapsed))
```

Line 32 prints the list `factors`, i.e., the list of the prime factors of the initial value of `n`. Line 33 stores the time elapsed during the program execution into the variable `elapsed`, which is printed on the terminal at line 34.

Thus, if you type, for instance `factorize.py 3512` at the terminal prompt, you get

```
$>factorize.py 3512
digits: 4
[2, 2, 2, 439]
elapsed time: 0.000120 s
```

the second line (first *output* line) tells you that you are factorizing a 4-digit integer, the third line gives you the actual factorization, $3512 = 2^3 \times 439$, and the last line tells you that the computer needed 1.2×10^{-4} s for computation. If you are using Spyder, the ooutput is written in the IPython console, the right lower subwindow of Fig. 1.2.

Obviously the factorization time is strongly computer-dependent! It is interesting to experiment with integers of different size on the same computer:

```
$>factorize.py 3891252771
digits: 10
```

```
[3, 3, 3, 7, 20588639]
elapsed time:  0.001096 s

$>factorize.py 348961235247715
digits: 15
[5, 131, 532765244653]
elapsed time:  0.118137 s

$>factorize.py 348961223452477159
digits: 18
[1831, 190585048308289]
elapsed time 1.443072 s

$>factorize.py 655449788001142878671
digits: 21
[655449788001142878671]
elapsed time: 3280.387602 s
```

Obviously the simple algorithm presented above can be improved, but the computation time of all known factorizing algorithms increases approximately exponentially with the number of digits of the integer to be factorized. For a given number of digits of n we observe the longest factorization time if n is prime. In the last example above 655 449 788 001 142 878 671 is a prime number.

The security of all current cryptographic algorithms relies on the fact that the factorization of large integers takes a very long time.

2.2 Functions

2.2.1 General

As in all programming languages, in Python a function is a block of code that performs a specific task, and that can be called several times by other parts of the program. Python comes with a huge amount of built-in (prewritten, precompiled and optimized) functions, some immediately available, and some packed into separate libraries, organized in *modules* and *packages*, that you must import into your code before use. We have already met the `import` command in several occasions. Some relevant mathematical functions available in Python are listed in Appendix A.

You can also write your own functions to perform specialized tasks. Unlike older high-level programming languages, and as in C/C++, in Python there is no distinction between *functions* and *subroutines*. The code of a Python function begins with a `def` statement and, if it must return values, ends with a `return` statement. The commands in the function definition are not executed as Python first passes over the lines, but only when the function is *called* by another part of the script. A function can have any number of arguments, including zero.

As a simple example, we can turn Script 2.1 into a function coded inside a larger script that checks which of the first 20 natural numbers are prime:

Listing 2.2 FuncFactorize.py

```
1  #!/usr/bin/env python3
2  import math
3  #
4  def factorize(n):
5    sqnf=int(math.ceil(math.sqrt(float(n))))
6    factors = []
7    while n%2==0:
8      factors.append(2)
9      n=n//2
10   i=3
11   while i<=sqnf:
12     while n%i==0:
13       factors.append(i)
14       n=n//i
15       sqnf=int(math.ceil(math.sqrt(float(n))))
16     i+=2
17   if n!=1:
18     factors.append(n)
19   return factors
20 #
```

Lines 4–19 define the function `factorize()`, which has the single argument `n` (the integer number to be factorized), and outputs the list of the factors of `n`. As stated above, a function can have any number of arguments, and the arguments are not restricted to belong to the same data type, they can be integers, floats, lists, strings, …, mixed in any possible way. In the present case we have a single integer argument, and the code of the function is identical to the corresponding code of Script 2.1. Line 19 returns the list of the prime factors of the argument `n`.

```
21 for i in range(2,21):
22   fact=factorize(i)
23   if len(fact)==1:
24     print(i,"is prime")
25   else:
26     print(i,fact)
```

The loop at Lines 21–26 is the main part of the script. Line 22 calls function `factorize()` with successive integer numbers `i` as arguments, and the factors of `i` are stored into the list `fact`. Line 23 checks if `fact` has only one element: in this case, `i` is prime, and this is stated by Line 24. Otherwise *i* and its prime factors are printed by line 26. The output of the script follows:

```
$>FuncFactorize.py
2 is prime
3 is prime
4 [2, 2]
5 is prime
6 [2, 3]
7 is prime
```

```
8 [2, 2, 2]
9 [3, 3]
10 [2, 5]
11 is prime
12 [2, 2, 3]
13 is prime
14 [2, 7]
15 [3, 5]
16 [2, 2, 2, 2]
17 is prime
18 [2, 3, 3]
19 is prime
20 [2, 2, 5]
```

2.2.2 Local and Global Variables

The variables of a function can be either *local* or *global*. All variables defined inside a function definition are local to the function by default. This means that whatever happens to this variable in the body of the function will have no effect on other variables of the same name outside of the function definition, if they exist. Thus, for instance, the variables `n`, `sqnf` and `factors` in Script 2.2 are local to the function `factorize()`. Line 22 copies the value of the variable `i`, local to the main body of the script, into the variable `n`, local to the function `factorize()`, and, after the function has been called, copies the output of the function (the list `factors`, local to `factorize()`), into the list `fact`, local to the main body of the script. Things can be made clearer by the simple examples discussed below.

In Listing 2.3 the function `func()` prints the square root of the variable x, that has not been defined anywhere.

Listing 2.3 GlobLoc0.py

```
#!/usr/bin/env python3
import math
#
def func():
  print(math.sqrt(x))
#
func()
```

Thus, Python complains when the script is called

```
$>GlobLoc0.py
Traceback (most recent call last):
  File "./GlobLoc0.py", line 7, in <module> func()
  File "./GlobLoc0.py", line 5, in func
    print(math.sqrt(x))
NameError: name 'x' is not defined
```

Python warns that Line 7 of the script calls the function `func()`, which prints the square root of variable x at Line 5, leading to an error because x has not been defined anywhere.

Listing 2.4 defines the variable x at Line 7, thus outside of the definition of `func()`.

Listing 2.4 GlobLoc1.py

```
1 #!/usr/bin/env python3
2 import math
3 #
4 def func():
5     print(math.sqrt(x))
6 #
7 x=5.0
8 func()
```

When Line 8 calls `func()`, Python interprets x as a *global* variable defined in the main body of the script because a variable named x local to `func()` does not exist. Thus, no error is found, and the square root of 5 is printed in the output

```
$>GlobLoc1.py
2.23606797749979
```

Finally, Script 2.5 defines a variable named x inside the function definition, at Line 5, and a further variable, also named x, in the main body of the script, at Line 8.

Listing 2.5 GlobLoc2.py

```
1 #!/usr/bin/env python3
2 import math
3 #
4 def func():
5     x=3.0
6     print(math.sqrt(x))
7 #
8 x=5.0
9 func()
10 print(math.sqrt(x))
```

Here two separate non-interacting variables coexist in the script, both variables are named x, but one of them is local to the main body of the script, while the other is local to function `func()`. The output, when the script is called, is

```
$>GlobLoc2.py
1.7320508075688772
2.23606797749979
```

Line 9 calls `func()`, which, at Line 6, prints the square root of its local variable x, i.e., $x = 3.0$, ignoring the variable x local to the main body of the script. Finally, Line 10, being outside of the function definition, ignores the variables local to the function, and prints the square root of the variable x local to the main body of the script, i.e., $x = 5.0$.

2.3 Reading and Writing Files

Up to now we have passed data to Python by typing values on the keyboard, thus through the *standard input*, and we read the results of Python elaborations on the computer monitor, or *standard output*. This is convenient for small tasks, but is absolutely not convenient for the elaboration of large data sets, as is almost always the case when we are dealing, for instance, with experimental results. When we must elaborate large amounts of data, it is convenient to read data stored in *input files* (this also avoids retyping all the data if our program crashes!), and have the computation results written into other, *output files*.

We start with a simple example. First, we write a text file containing one number per line with our favorite editor, and name it `numdata.txt`. For instance:

```
2.0
4.0
6.0
8.0
10.0
12.0
14.0
16.0
18.0
20.0
```

The following simple script reads `numdata.txt` and displays its content on the terminal

Listing 2.6 ReadFile0.py

```
1 #!/usr/bin/env python3
2 #
3 hnd=open("numdata.txt","r")
4 num=hnd.readlines()
5 print(num)
6 hnd.close()
```

Line 3 opens the file in *read mode*. The name of the file, being a string, must be within quotes. The argument `"r"` stands for *read mode*, meaning that the script can *read*, but not in any way *modify*, the file. The variable `hnd` is the *file handler*, a pointer to the file in the computer memory. All successive operations on the file go through this handler. Line 4 creates the list `num`, where the method `hnd.readlines()` copies all the lines of the file. Line 6 *closes* the file, so that the file is no longer accessible to the script. It is very important to always close a file that has been opened, in order to avoid that it may be corrupted. This is what we see when we run the script

```
>ReadFile0.py
['2.0\n', '4.0\n', '6.0\n', '8.0\n', '10.0\n', '12.0\n', '14.0\n', '16.0\n',
'18.0\n', '20.0\n', '\n']
```

All items of the list are within quotes because they are interpreted as *strings*, not as *numbers*, and all of them end with a newline sign `\n` because we wrote one number per line. The last `\n` means that we typed a blank line at the end of the file. If we want to elaborate these numbers, we must first convert them to regular floats. The script that follows does the job

Listing 2.7 ReadFile.py

```
1  #!/usr/bin/env python3
2  from math import sqrt
3  #
4  num=[]
5  hnd=open("numdata.txt","r")
6  while True:
7      buff=hnd.readline()
8      if not buff:
9          break
10     try:
11         num.append(float(buff.strip()))
12     except ValueError:
13         pass
14 hnd.close()
15 #
16 hnd=open("sqrtdata.txt","w")
17 for i in range(len(num)):
18     hnd.write('{:6.2f}{:12.4f}\n'.format(num[i],sqrt(num[i])))
19 hnd.close()
```

Line 2 imports the function `sqrt()` from `math`, we shall need it for evaluating square roots. Line 4 creates the list `num` where we shall store the numbers read from the file. Line 5 opens the file in read mode. Lines 6–13 define an *infinite* loop for reading the file. We use an infinite loop because we don't know the length of the file before reading it. At each loop iteration, Line 7 copies successive lines of the file into the list `buff`. Note that, while the method `readlines()` encountered in Script 2.6 reads all the lines of the file in a single step, the method `readline()` reads a single line at a time. When the end of the file is reached, `readline()` returns an empty string, which is considered as equivalent to *False* by the `if` statement of line 8. Thus, the loop is interrupted when the end of the file is reached. Lines 10–13 check if it is possible to convert each line into a float number. The method `strip()` strips all blank and newline characters from the beginning and end of each string (thus, for instance, the newline characters `\n` at the end of each line are *stripped*). What remains of each string is converted to a float and appended to the list `num`. If the conversion to float fails (for instance, there is nothing to convert at the last line, which comprises only a `\n` character), a *ValueError* exception is raised, and the `try...except` statement simply skips the action, and nothing is appended to `num`. Line 14 closes the input file, and frees the file handler `hnd`. Line 16 opens an output file named `sqrtdata.txt` in *write mode* (argument `"w"`). *Write mode* means that a file named `sqrtdata.txt` will be created on the disk if it does not exist already. If a file of that name exists, it will be overwritten, and its previous content will be permanently lost. The loop at lines 16–17 writes lines one by one into the output file, each line comprising a number of the input file and its square root. Finally, line 19 closes the output file. When you run Script 2.7 you don't see anything on the terminal, but a file named `sqrtdata.txt` is created, and you can read it with an editor. This is its content

```
 2.00    1.4142
 4.00    2.0000
 6.00    2.4495
 8.00    2.8284
10.00    3.1623
12.00    3.4641
14.00    3.7417
16.00    4.0000
18.00    4.2426
20.00    4.4721
```

2.4 Calling External Commands from Python Scripts

It is possible to call external commands (for instance operating-system commands) from a Python script, and to write Python scripts more or less equivalent to Unix Bash files and Windows batch files. This can be done through methods of the *os* and *subprocess* modules. For instance, line 3 of Listing 2.8 calls the VLC media player, available for Linux, Windows and Mac OSX.

Listing 2.8 Calling VLC Media Player

```
1 #!/usr/bin/env python3
2 import os
3 os.system('vlc')
```

The external command `vlc` is inserted into a string, which is passed as argument to the method `os.system()`.

It is often necessary to pass parameters to an external command. This is done by inserting the parameters into the string argument. For instance, under Linux, when you wish the list of the contents of a directory, you can type `ls -latr` in a terminal. Here, `ls` is the command that lists the directory contents. The meanings of the four letters in the option string are: `l`: use a long listing format, `a`: include entries starting with "." ("hidden" files), `t`: sort by modification time, newest first, `r`: reverse order while sorting (i.e., newest last). The result is similar to what you obtain by typing the command `dir` in a Windows terminal.

Listing 2.9 Listing the contents of a directory

```
#!/usr/bin/env python3
import os
os.system('ls -latr')
```

The `os.system()` method can also launch multiple commands separately. This is done by writing each command, with its parameters, in a separate string, and then passing the string commands, joined by vertical bars, "|", as argument to `os.system()`.

Listing 2.10 Calling VLC and Listing a Directory

```
#!/usr/bin/env python3
import os
os.system('ls␣-latr' | 'vlc')
```

Thus, Listing 2.10 simultaneously lists the contents of the directory and calls the *VLC* media player. Obviously this is of no practical interest whatsoever here, but we shall see later on that this trick can be used, for instance, for performing calculations in parallel on multi-processor computers.

2.5 Exercises

Exercise 2.1 Write a script that takes as command-line argument an integer number and prints its factorial.

Exercise 2.2 Write a script that takes as command-line arguments the x and y components of a 2D vector, v_x and v_y, and an angle α in degrees. The script outputs the two components of the vector after a rotation by α.

Exercise 2.3 Write a script that prints the list of the factorials of the integers in the range 0–10.

Exercise 2.4 Write a script that generates a file containing the list of the integers in the range 0–20 with their respective square and cube roots, each with 6 digits after the decimal point.

Exercise 2.5 In mathematics, Pascal's triangle, also known as Tartaglia's triangle, is a triangular array of the binomial coefficients, i.e., the coefficients for the expansions of powers of a binomial, $(a+b)^n$. The rows of Pascal's triangle are conventionally enumerated starting with row $n = 0$ at the top (the 0th row). The triangle may be constructed in the following manner: In row 0 there is a unique nonzero entry, namely 1. Each entry of each subsequent row is constructed by adding the number above and to the left with the number above and to the right, treating blank entries as 0. The first 5 rows are thus

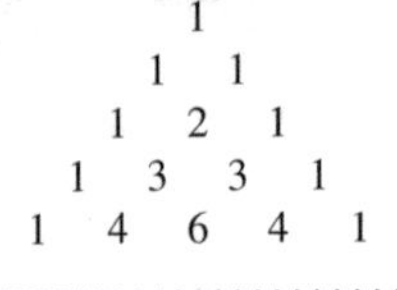

Write a script that prints the first 10 rows of Pascal's triangle.

Chapter 3
Plotting with Matplotlib

3.1 Pyplot

Pyplot is a collection of command-style functions that make Matplotlib work like MATLAB. Each pyplot function makes some change to a figure: e.g., creates a figure, creates a plotting area in a figure, plots lines in a plotting area, adds labels (text) to the plot itself or to the horizontal and vertical axes of the figure, … Successive calls to different *pyplot* functions preserve what done in the figure by the previously called functions, so that the complete figure can be drawn in successive steps.

3.2 Plotting Lists of Numbers

As a first, simple example let us see how we can plot a single list of numbers, for instance the list [1.0, 2.0, 4.0, 8.0, 16.0, 32.0, 64.0]. We can do this by writing the Python script shown in Listing 3.1

Listing 3.1 Plot1List.py

```
1  #!/usr/bin/env python3
2  import matplotlib.pyplot as plt
3  #
4  plt.plot([1.0,2.0,4.0,8.0,16.0,32.0,64.0])
5  plt.show()
```

Line 2 imports `matplotlib.pyplot` as `plt`. The function `plot()` is called at Line 4, and generates the polyline shown in Fig. 3.1. Note that, if we provide a single list, or array, of numbers to the `plot()` function, *matplotlib* assumes that we are giving a sequence of *ordinates* (y values), and the corresponding abscissae (x values) are generated automatically. The default list of x values has the same length as the list

---

Electronic supplementary material The online version of this chapter (https://doi.org/10.1007/978-3-030-45027-4_3) contains supplementary material, which is available to authorized users.

G. Moruzzi, *Essential Python for the Physicist*,
https://doi.org/10.1007/978-3-030-45027-4_3

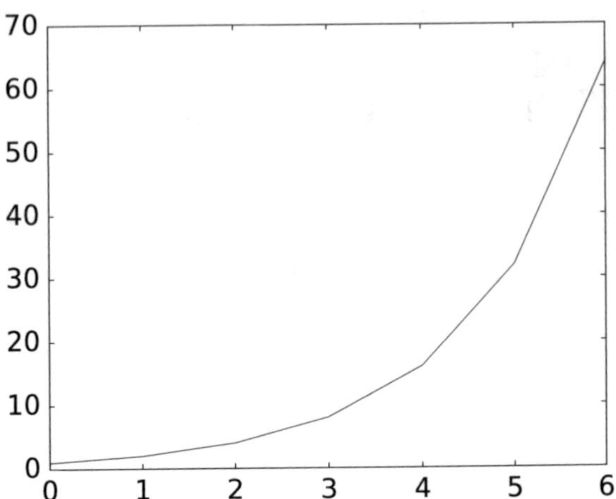

Fig. 3.1 Plot generated by Listing 3.1

of y values, and comprises successive natural numbers starting with 0. In the present case, this leads to the correct plot, since we are actually plotting $y = 2^x$. In the more general case, we pass `plot()` two lists of equal length, the first of which will be considered as the list of the *abscissae*, and the second as the list of the ordinates of the points to be plotted. The figure will be automatically scaled according to the input x and y values. We can also pass an optional third argument which is a *format string* indicating the color and line type of the plot. The default format string is '`b-`', which generates a solid blue line as in Fig. 3.1. Listing 3.2 gives a further example (Fig. 3.2).

Listing 3.2 Plot2Lists.py

```
1 #!/usr/bin/env python3
2 import matplotlib.pyplot as plt
3 #
4 plt.plot([3.0,4.0,5.0,6.0,7.0,8.0],[8.0,16.0,32.0,64.0,128.0,256.0],'ro')
5 plt.show()
```

Line 4 passes the two x and y lists to `plot()`, while the third argument, the string '`ro`', asks for the plot to be represented by red (`r`) circles (`o`). The default value of this argument is '`b-`', corresponding to a blue (`b`) continuous line (`-`), as in the case of Fig. 3.1. *Matplotlib* is not limited to working with lists: this would be definitely uncomfortable for numeric processing. Rather, we shall normally use *numpy* arrays instead of hand-written lists. In fact, all sequences are converted to numpy arrays internally. Listing 3.3 plots three different lines, with different format styles, in a single command using arrays (Fig. 3.3).

Listing 3.3 multiplot.py

```
#!/usr/bin/env python3
import matplotlib.pyplot as plt
import numpy as np
#
x=np.arange(0.0,2.0,0.05)
plt.plot(x,x,x,np.sqrt(x),'ro',x,x**2,'g^')
plt.show()
```

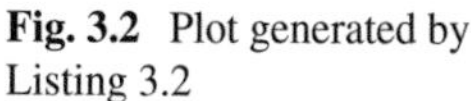

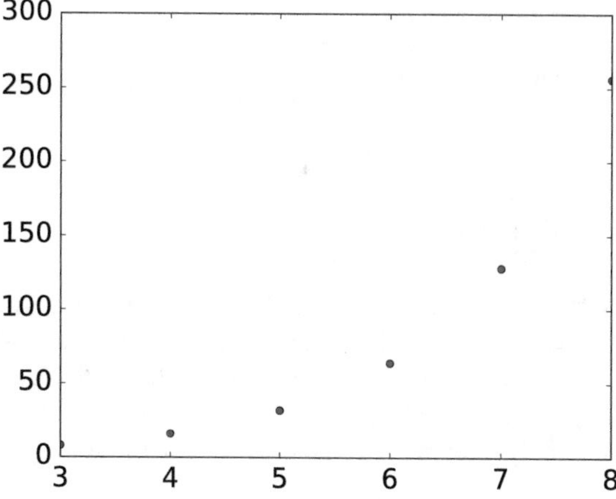

Fig. 3.2 Plot generated by Listing 3.2

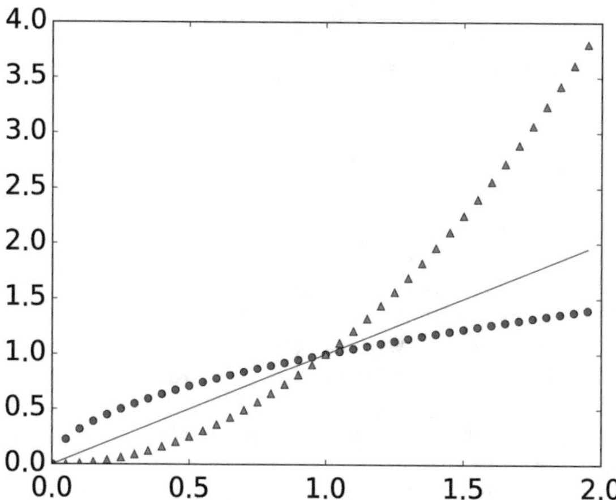

Fig. 3.3 Superposed plots

Line 5 creates the array `x` comprising the 40 elements 0.0, 0.05, 0.1, ..., 1.95. The first two arguments of the `plot()` function at line 6 plot the array $y = x$ as a function of x, thus resulting in a straight line. No format string is given, consequently the default blue continuous line is used for the plot. The three following arguments, `x`, `np.sqrt(x)` and the format string '`ro`', plot $y = \sqrt{x}$ as a function of x, using red circles. Finally, the last three arguments `x`, `x**2` and the string '`g^`', plot $y = x^2$ versus x using green triangles. The three plots are superposed in Fig. 3.3.

3.3 Plotting Functions

We can plot all the functions supplied by, for instance, the *numpy* and *scipy* packages, as well as any user-defined function, by first sampling x and y coordinates into two `numpy` arrays, and then passing the arrays to `plot()`. As a first example we plot a sine curve using Listing 3.4

Listing 3.4 PlotSin.py

```
1 #!/usr/bin/env python3
2 import numpy as np
3 import matplotlib.pyplot as plt
4 #
5 x = np.arange(0,6.4,0.1);
6 y = np.sin(x)
7 plt.plot(x, y)
8 plt.show()
```

Line 2 imports `numpy` as `np`. Line 5 creates the `x` array, comprising the 64 numbers 0, 0.1, 0.2, ..., 6.3. Line 6 creates the `y` array, comprising the 64 values $y_i = \sin(x_i)$. Line 7 creates the plot of y versus x, and line 8 displays the plot on the computer monitor, as shown in Fig. 3.4. The plot can be refined by calling further *pyplot* functions. A slightly modified version of Listing 3.4 follows

Listing 3.5 PlotSinFig2.py

```
#!/usr/bin/env python3
import numpy as np
import matplotlib.pyplot as plt
#

x = np.arange(0,6.4,0.1);
y = np.sin(x)
plt.plot(x, y)
plt.grid(True)
plt.xticks(fontsize=14)
plt.yticks(fontsize=14)
plt.ylabel(r'$\sin(x)$',fontsize=24)
plt.xlabel(r'$x/{\rm rad}$',fontsize=24)
plt.tight_layout()
plt.savefig('SineGrid.eps',format='eps',\
   dpi=1000)
plt.show()
```

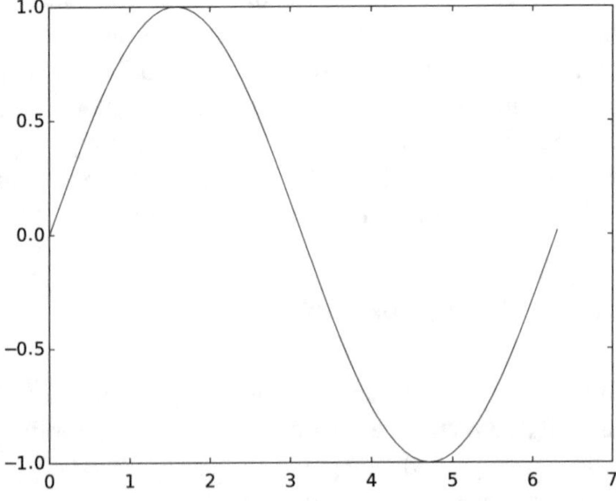

Fig. 3.4 Plot generated by Listing 3.4

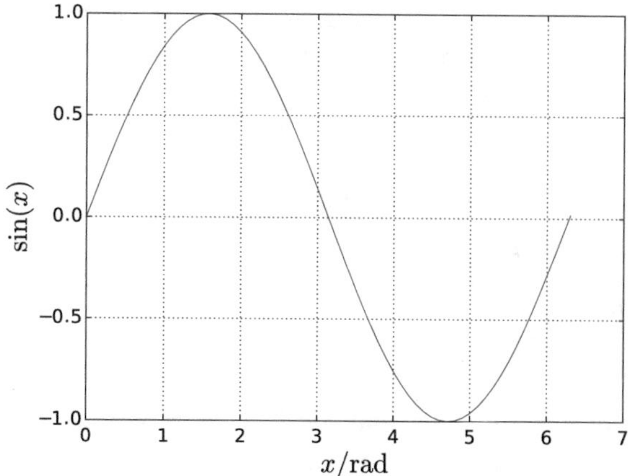

Fig. 3.5 Plot generated by Listing 3.5

Line 8 adds a grid to the plot, Lines 9 and 10 change the default font sizes of the x and y scales of the grid, respectively. Lines 11 and 12 write labels for the y and x axes of the plot, respectively. The labels must be enclosed in single or double quotes ("or"). Labels can be in plain text, but LaTeX text, enclosed between $ signs, can also be used. In this case the text string must be preceded by an r. The leading r is important: it signifies that the string is a *raw string*, where, for instance, backslashes must not be interpreted as python escapes. Again, the font size of a label can be specified. It is advisable to experiment a little bit with different font sizes at lines 9–13 in order to obtain an aesthetically satisfactory result. Line 13 adjusts the layout of the figure in order to include the grid ticks and labels: as an experiment, see what happens if you move line 13 immediately after line 8, or 9. Line 14–15 creates an Encapsulated PostScript figure named 'SineGrid.eps'. The supported formats for the output figure are *.png*, *.pdf*, *.ps*, *.eps* and *.svg*. The result is shown in Fig. 3.5.

3.4 Multiple Figures

A single Python script can generate a single figure comprising more than one separate plots (*subplots*). An example is shown in Listing 3.6

Listing 3.6 SubPlots.py

```
#!/usr/bin/env python3
import matplotlib.pyplot as plt
import numpy as np

def fun1(t,omega,tau,ampl):
    y=ampl*np.cos(omega*t)*np.exp(-t/tau)
    return y
def fun2(t,omega,omega2,ampl):
    y=ampl*np.sin(omega2*t)*np.sin(omega*t)
    return y
#
```

Lines 5–7 and 8–10 define the two functions `fun1(t,omega,tau,ampl)` and `fun2(t,omega, omega2,ampl)`, in mathematical form they are

$$f_1(t, \omega, \tau, A) = A \cos(\omega t)\, \mathrm{e}^{-t/\tau} \quad \text{and} \quad f_2(t, \omega, \omega_2, B) = B \sin(\omega_2 t) \sin(\omega t), \tag{3.1}$$

respectively. These two functions will be plotted in two separate subplots.

```
12 omega=32.0
13 omega2=np.pi/2.0
14 tau=1.0
15 ampl=5.0
16 ampl2=10.0
```

Lines 12–16 assign numerical values to parameters of the two functions f_1 and f_2 of (3.1), namely $\omega = 32$, $\omega_2 = \pi/2$, $\tau = 1.0$, $A = 5.0$ and $B = 10.0$.

```
17 t=np.arange(0.0, 2.0, 0.01)
18 s1=fun1(t,omega,tau,ampl)
19 s2=fun2(t,omega,omega2,ampl2)
20 # ........................................
21 plt.figure(figsize=(10,4))
```

Line 17 creates the array `t`, comprising the 200 numbers [0.00, 0.001, 0.002, ..., 0.199] that will be used as abscissae for the plots. Lines 18 and 19 create the two arrays `s1` and `s2`, used as ordinates in the plots. Line 21 creates a Fig. 3.10 inches wide and 4 inches high. The absolute values (or units) are actually not relevant, since the figure will be scaled both on the computer monitor and in printing, but the width/height ratio **is** relevant.

```
22 # .................................. subplot 1
23 plt.subplot(1,2,1)
24 plt.plot(t,s1)
25 plt.plot(t,ampl*np.exp(-t/tau))
26 plt.grid(True)
27 plt.xticks(fontsize=14)
28 plt.yticks(fontsize=14)
29 plt.ylabel(r'$A\,\cos(\omega t)\,{\rm e}^{-t/\tau}$',fontsize=24)
30 plt.xlabel(r'$t$',fontsize=24)
31 plt.tight_layout()
```

Line 23 creates the first subplot. The arguments 1, 2, 1 of the function `subplot()` assign 1 row and 2 columns to the complete figure, thus the two subplots will be located side by side on the same row. The last argument, 1, tells that what follows will be drawn in the first (left) subplot. Line 24 plots the array `s1` versus the array `t`, i.e., f_1 versus t. Line 25 adds the plot of the exponential $A\,\mathrm{e}^{-t/\tau}$ versus t. The rest is analogous to lines 8–13 of Listing 3.5. The two superposed plots are shown in the left part of Fig. 3.6

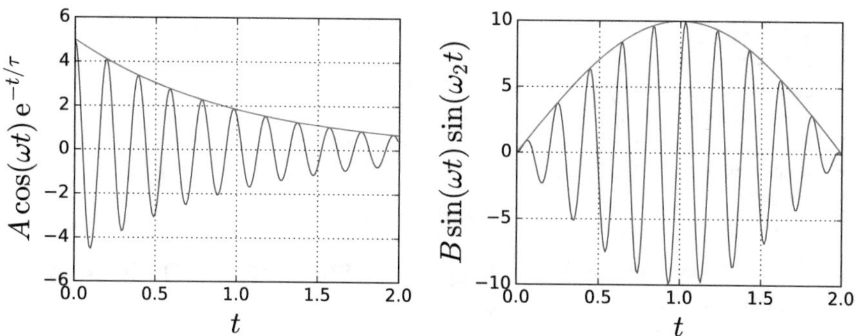

Fig. 3.6 The two subplots generated by Listing 3.6

```
32 # .................................. subplot 2
33 plt.subplot(1,2,2)
34 plt.plot(t,s2)
35 plt.plot(t,ampl2*np.sin(omega2*t))
36 plt.grid(True)
37 plt.xticks(fontsize=14)
38 plt.yticks(fontsize=14)
39 plt.ylabel(r'$B\,\sin(\omega t)\,\sin(\omega_2 t)$',fontsize=24)
40 plt.xlabel(r'$t$',fontsize=24)
41 plt.tight_layout()
42 #
```

Line 33 creates the second subplot, located on the second column of the first (and only) row of the figure, i.e., at the right. Line 34 plots the array `s2` versus the array `t`, i.e., f_2 versus t. Line 35 adds the plot of the low-frequency sine $B\sin(\omega_2 t)$ versus t. Lines 36–41 add grid and axis labels, and adjust the font sizes. The plot is shown in the right part of Fig. 3.6.

```
43 plt.savefig('MultiPlot0.eps',format='eps',dpi=1000)
44 plt.show()
```

A figure in Encapsulated PostScript is created, and the two plots are shown on the computer monitor.

3.5 Logarithmic Axis Scales

Pyplot supports not only linear axis scales, but also *logarithmic* and *logit* scales. This can be useful when data span many orders of magnitude. Listing 3.7 gives an example.

Listing 3.7 LogPlot.py

```
1  #!/usr/bin/env python3
2  import matplotlib.pyplot as plt
3  import numpy as np
4  # .................................. functions
5  x=np.arange(1.0,20.5,1.0)
6  y1=x**2
7  y2=np.sqrt(x)
```

Line 5 creates the array x= [1.0, 2.0, ..., 20.0]. Lines 6 and 7 create the arrays y1 and y2, with $y_i^{(1)} = x_i^2$ and $y_i^{(2)} = \sqrt{x_i}$.

```
8  # ........................................
9  plt.figure(figsize=(10,8))
```

Analogous to Line 21 of Listing 3.6. Again, only the width/height ratio of the whole figure is relevant to us.

```
10 # .................................. subplot 1
11 plt.subplot(2,2,1)
12 plt.plot(x,y1,'ro')
13 plt.plot(x,y2,'bo')
14 plt.grid(True)
15 plt.xticks(fontsize=14)
16 plt.yticks(fontsize=14)
17 plt.xlabel('linear',fontsize=16)
18 plt.ylabel('linear',fontsize=16)
19 plt.tight_layout()
20 plt.ylim(-10,420)
21 plt.xlim(0,21)
```

This is the first of four sublopts, located in the first position, (1, 1) or upper left, of our 2 × 2 plot array shown in Fig. 3.7. All four plots display the same data, i.e., y1 versus x superposed to y2 versus x. The y1 data are represented by red circles, the y2 data by blue circles in all plots. In this first plot both the x and the y scales are linear, and the plot of the square root data results very flat. Lines 20 and 21 set the limits of the x and y values, slightly extending their default values. This is done in order to keep the circles representing the extreme data points within the plot frame. As an experiment, see what happens if you comment out these lines.

```
22 # .................................. subplot 2
23 plt.subplot(2,2,2)
24 plt.plot(x,y1,'ro')
25 plt.plot(x,y2,'bo')
26 plt.grid(True)
27 plt.yscale('log')
28 plt.xticks(fontsize=14)
29 plt.yticks(fontsize=14)
30 plt.xlabel('linear',fontsize=16)
31 plt.ylabel(r'$\log$',fontsize=20)
32 plt.tight_layout()
33 plt.ylim(0.8,500)
34 plt.xlim(-0.8,21)
```

This is the second (upper right) subplot. Line 27 makes the y scale logarithmic, with this choice the plot of the square roots is no longer flat. Remember that, when you use a logarithmic scale, negative and zero coordinate values must be avoided. In case you need negative coordinate values, Line 27 can be replaced by '`plt.yscale(symlog)`', where '`symlog`' stands for *symmetric logarithm*. The function symlog() is defined as

$$\text{symlog}(y) = \begin{cases} -\log(-y) & \text{if} \quad y < -y_{\text{thr}} \\ \alpha y & \text{if} \quad -y_{\text{thr}} \leq y \leq y_{\text{thr}} \\ \log(y) & \text{if} \quad y > y_{\text{thr}} \end{cases} \tag{3.2}$$

so that it is linear in the range $-y_{\text{thr}} \leq y \leq y_{\text{thr}}$, avoiding the logarithm singularity at $y = 0$. By default $y_{\text{thr}} = 2$, but you can change this value to, for instance, 0.1, by modifying Line 27 as follow as

```
27  plt.yscale('symlog',linthreshy=0.1).
```

The constant α of (3.2) is defined as $\alpha = \log(y_{\text{thr}})/y_{\text{thr}}$, so that symlog($y$) is continuous at $y = \pm y_{\text{thr}}$. See Exercise 3.3.

```
35  # .................................. subplot 3
36  plt.subplot(2,2,3)
37  plt.plot(x,y1,'ro')
38  plt.plot(x,y2,'bo')
39  plt.grid(True)
40  plt.xscale('log')
41  plt.xticks(fontsize=14)
42  plt.yticks(fontsize=14)
43  plt.xlabel(r'$\log$',fontsize=20)
44  plt.ylabel('linear',fontsize=16)
45  plt.tight_layout()
46  plt.ylim(-10.8,420)
47  plt.xlim(0.8,22)
```

This is the third (lower left) subplot. Line 40 makes the x scale logarithmic, while the y scale is linear. In the case of zero or negative x values, the logarithm can be replaced by the function symlog() defined by (3.2) by the command `plt.xscale('symlog',linthreshx=MyValue)`.

```
48  # .................................. subplot 4
49  plt.subplot(2,2,4)
50  plt.plot(x,y1,'ro')
51  plt.plot(x,y2,'bo')
52  plt.grid(True)
53  plt.xscale('log')
54  plt.yscale('log')
55  plt.xticks(fontsize=14)
56  plt.yticks(fontsize=14)
57  plt.xlabel(r'$\log$',fontsize=20)
58  plt.ylabel(r'$\log$',fontsize=20)
59  plt.tight_layout()
60  plt.ylim(0.8,500)
61  plt.xlim(0.8,25)
```

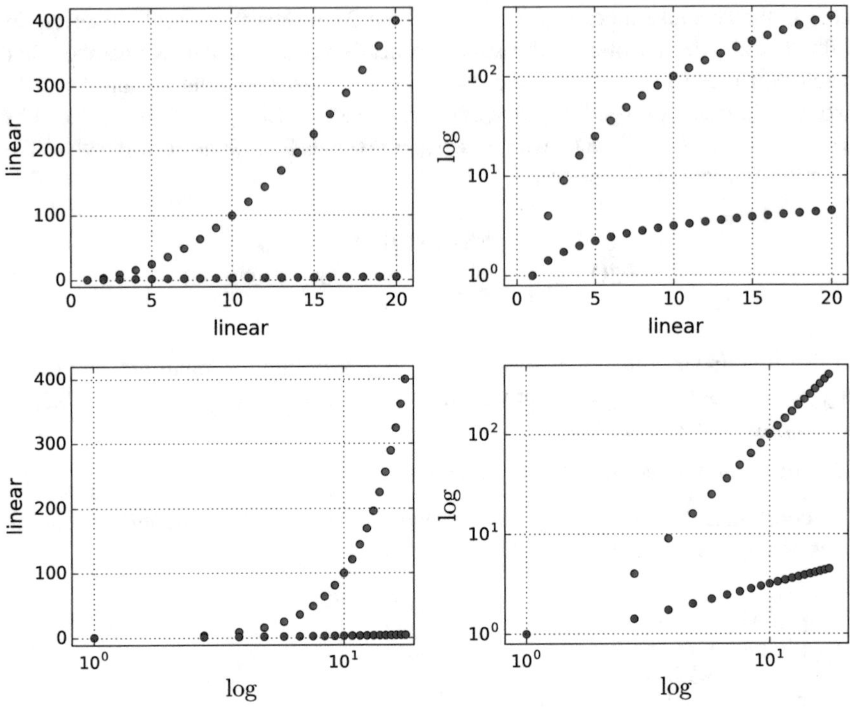

Fig. 3.7 The four subplots generated by Listing 3.7

This is the fourth (lower right) subplot. Lines 53 and 54 make the scales of both axes logarithmic. The circles representing both data sets lie on straight lines, because we have

$$\log y^{(1)} = \log x^2 = 2 \log x \qquad \text{and} \qquad \log y^{(2)} = \log \sqrt{x} = \frac{1}{2} \log x. \tag{3.3}$$

```
62  #
63  plt.savefig('LogPlot0.eps',format='eps',dpi=1000)
64  plt.show()
```

3.6 Plotting Field Lines

In this section we consider how to plot field lines, specifically how to plot the field lines of an electric dipole. Our dipole consists of two charges, $+q$ and $-q$, located at $(0, +h)$ and $(0, -h)$ in a Cartesian reference frame, respectively, as shown in Fig. 3.8. Extensions to the field lines generated by more complicated charge configurations

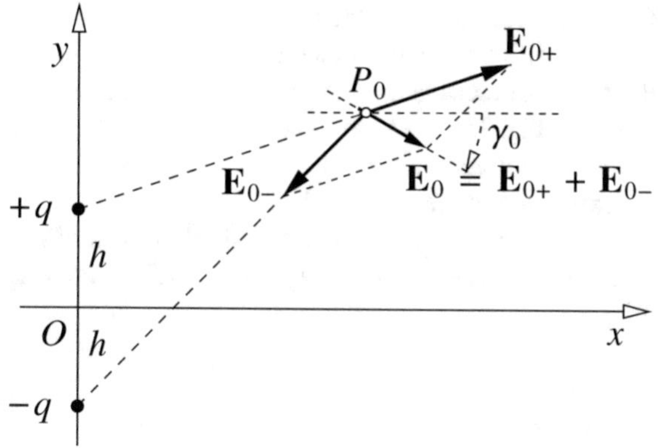

Fig. 3.8 The Electric field generated by the two charges $\pm q$ in P_0

should be straightforward. The idea is using a method analogous to the Euler method for solving ordinary differential equations, discussed later on in Sect. 5.2. In other words, we approximate each field line by a polyline comprising very short line segments of equal length. Each polyline starts from an initial point $P_0 \equiv (x_0, y_0)$, close to, but obviously not coinciding with, one of the two charges $\pm q$ generating the field. Once chosen the start point P_0 of a field line, we evaluate the electric field $\boldsymbol{E}_0$ in it. What is relevant for our method is not the intensity of the field, but only its direction, which forms an angle $\gamma_0 = \arctan(E_{0y}/E_{0x})$, negative in the case of Fig. 3.8, with the x direction. Once determined the angle γ_0, we choose a small length $\delta\ell$, which will be the common length of all the single line segments forming the polyline approximating the field line. In general, some experimenting on the most convenient value for $\delta\ell$ can be needed. The second vertex of our polyline is the point $P_1 \equiv (x_1, y_1) = (x_0 + \delta\ell \cos\gamma_0, y_0 + \delta\ell \sin\gamma_0)$. The next step is the evaluation of the electric field $\boldsymbol{E}_1$ in P_1 and its angle γ_1 with the x direction, which determines the point P_2. The procedure is iterated for finding the successive vertices P_m of the polyline. We shall stop when the polyline gets to close to the other point charge, or when it exits our intended plotting area. A possible procedure is shown in Listing 3.8.

Listing 3.8 PlotDipoleField.py

```
#!/usr/bin/env python3
import matplotlib.pyplot as plt
import numpy as np
# ............................................ charge locations
yplus=1.0
yminus=-1.0
xplus=xminus=0.0
rad=0.1
lim=30
nLin=60
plt.axis('off')
```

Quantities `xplus` (`xminus`) and `yplus` (`yminus`) are the x and y coordinates of the positive (negative) charge of the dipole, in arbitrary units. Quantity `rad` is the radius of a circle around the positive charge $+q$, from where our field lines will start, see Fig. 3.9. We stop the evaluation of a field line when the x, or y coordinate of the current vertex is greater than `lim`, or smaller than `-lim`. Both `rad` and `lim` are expressed in the same arbitrary units as `xplus`, `xminus`, `yplus` and `yminus`. Quantity `nLin` is the number n of field lines starting from the positive charge. Line 11 removes the axes from the figure. You can comment this line out, and see what happens.

```
12 # ...................................................... field lines
13 i=0
14 delta=0.05
```

Variable `i` is a counter for the field lines, while `delta` is the length $\delta\ell$ of the line segments.

```
15 while i<nLin:
16   xlist=[]
17   ylist=[]
18   alpha0=i*2*np.pi/float(nLin)
19   x=xplus+rad*np.cos(alpha0)
20   y=yplus+rad*np.sin(alpha0)
21   xlist.append(x)
22   ylist.append(y)
23   while True:
24     alpha=np.arctan2((y-yplus),(x-xplus))
25     beta=np.arctan2((y-yminus),(x-xminus))
26     Eplus=1.0/((x-xplus)**2+(y-yplus)**2)
27     Eminus=-1.0/((x-xminus)**2+(y-yminus)**2)
28     Ex=Eplus*np.cos(alpha)+Eminus*np.cos(beta)
29     Ey=Eplus*np.sin(alpha)+Eminus*np.sin(beta)
30     gamma=np.arctan2(Ey,Ex)
31     x=x+delta*np.cos(gamma)
32     y=y+delta*np.sin(gamma)
33     if x>lim or x<-lim or y>lim or y<0:
34       break
35     xlist.append(x)
36     ylist.append(y)
37   plt.plot(xlist,ylist,'k-',linewidth=0.5)
38   ylist=-np.array(ylist)
39   plt.plot(xlist,ylist,'k-',linewidth=0.5)
40   i+=1
```

Loop 15–40 draws a single field line at each iteration. Lines 16 and 17 create the (initially empty) lists of the x and y coordinates of the vertices of the approximating polyline. Angle `alpha0`, $\alpha_0 = 2\pi i/n$ (n being the number of plotted field lines), defined at Line 18 and shown in Fig. 3.9, determines the position of the start point P_0 of the current field line. All field lines start from points equally spaced on a small circumference of radius `rad` (r in the figure) centered around the positive charge $+q$. The x and y coordinates of P_0 are evaluated at lines 19 and 20, respectively, and inserted into `xlist` and `ylist` at lines 21 and 22. The loop 23–36 evaluates the

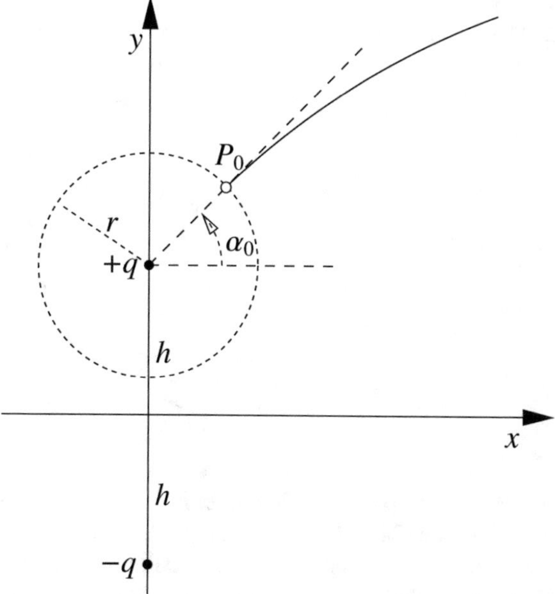

Fig. 3.9 The starting points of the field lines are located on a circleof radius r around the positive charge.

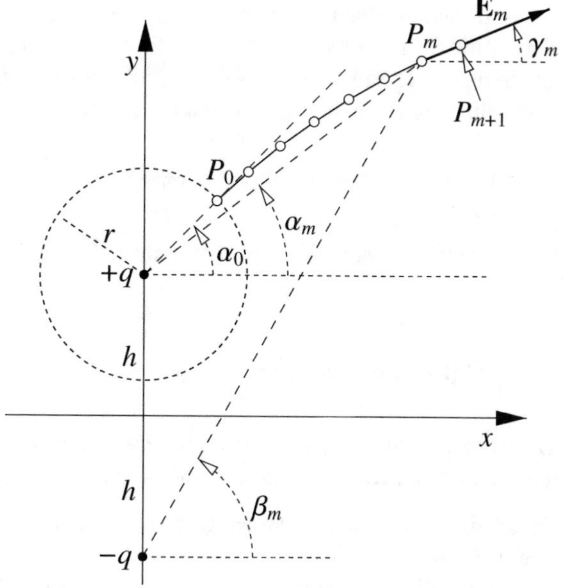

Fig. 3.10 Evaluation of the elctric field at the successive vertices of the polyline approximating a field line of an electric dipole

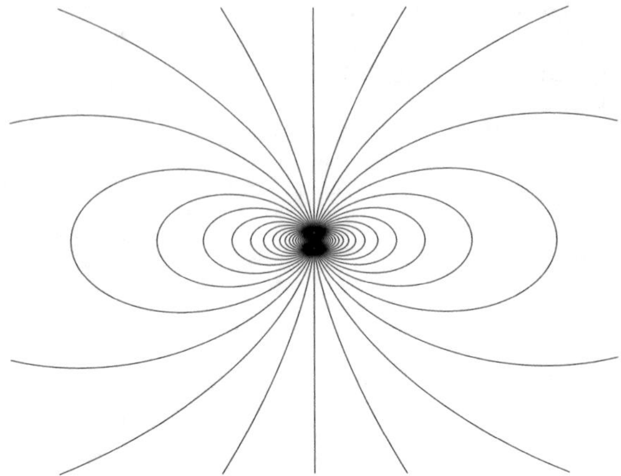

Fig. 3.11 Final plot of the field lines of an electric dipole

successive vertices of the field line, P_m, as shown in Fig. 3.10. Variables `alpha` and `beta` at Lines 24 and 25 are the angles α_m and β_m of Fig. 3.10. Lines 26–29 evaluate the x and y components of the electric field $\boldsymbol{E}_m$ at P_m, and Line 30 the angle γ_m that $\boldsymbol{E}_m$ forms with the x axis. Lines 31 and 32 evaluate the coordinates of the successive point of the "field polyline", P_{m+1}. Lines 33 and 34 break the loop if either the x or the y coordinate is out of range. Only positive y coordinates are accepted, since we can exploit the mirror symmetry of the figure about the x axis. Line 37 plots the field line in the upper half-plane $y > 0$. Line 38 changes the signs of `ylist`, and line 39 draws the symmetric field line in the lower half-plane.

```
41  # ..............................................................
42  plt.savefig('DipoleField.pdf',format='pdf')
43  plt.show()
```

Line 42 saves the plot into a pdf file named DipoleField.pdf, shown in Fig. 3.11.

3.7 Pyplot Animation

The *matplotlib* library provides the possibility to produce animations on the computer monitor. The easiest way is to use one of the following two animation classes

FuncAnimation: makes an animation by repeatedly calling a provided function `func()`.

ArtistAnimation: this animation uses a fixed set of Artist objects.

In the present context we shall consider the class *FuncAnimation* only, which is of more interest for a physicist. As a simple example, we shall write a script displaying a horizontally translating (panning) plot of an exponentially damped sine curve.

Listing 3.9 SineDecay.py

```
1  #!/usr/bin/env python3
2  #
3  import numpy as np
4  import matplotlib.pyplot as plt
5  import matplotlib.animation as animation
6  #
7  fig,ax=plt.subplots(figsize=(10,6))
8  ax.set(ylim=(-1,1))
9  #
```

Line 7 creates a figure of sizes 10×6 inches with a single subplot named `ax`. Line 8 sets the limits of the y axis of the subplot. This is needed to prevent `pyplot` from rescaling the vertical axis of the plot when the amplitude of the plotted function diminishes.

```
10 def func(x):
11     return np.sin(8*x)*np.exp(-0.1*x)
12 #
13 x=np.arange(0,10,0.01)
14 line,=ax.plot(x,func(x))
15 #
```

Lines 10–11 define the function $f(x) = \sin(8x)\,\mathrm{e}^{-0.1x}$, that we are going to plot. Line 13 stores the initial 100 x values at which the function is to be evaluated, namely $[0, 0.01, 0.02, \ldots, 9.99]$, into the list `x`. Line 14 creates the object `line`, belonging to class *matplotlib.lines.Line2D*, in the subplot `ax`. Object `line` has the list `x` as *xdata*, and the list `func(x)` as *ydata*.

```
16 def animate(i):
17     xx=x+i/100
18     line.set_xdata(xx)
19     line.set_ydata(func(xx))
20     ax.set(xlim=(xx[0],xx[999]))
21     return line,
22 #
```

Lines 16–21 define the function `animate()`, which updates the plot at each animation "frame". The function has the single argument `i`, which will be increased by 1 at each animation step. Line 17 creates the list `xx`, whose elements are obtained from the corresponding elements of the original list `x` by adding `i/100` to each of them. Lines 18 and 19 update the x and y values of the object `line`, respectively. Line 20 updates the limits of the x axis of the plot. Line 21 returns the object `line`.

```
23 ani=animation.FuncAnimation(fig,animate,interval=20)
24 plt.show()
```

At line 23, the function `FuncAnimation()` creates the plot animation by repeatedly calling the function `animate()`. The mandatory arguments of `FuncAnimation()` are two: the figure where to draw, `fig` in our case, and the updating function, here `animate`. Function `FuncAnimation()` has many optional further arguments, here we use only one, namely `interval`, the delay

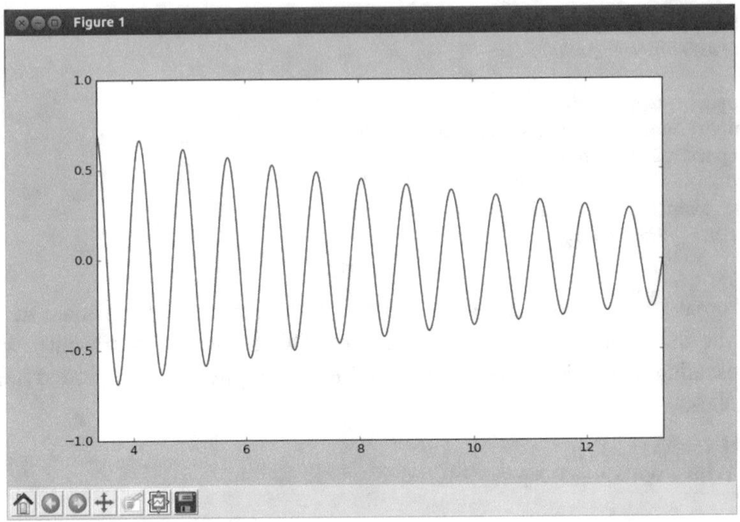

Fig. 3.12 The decaying sine wave of Listing 3.9

time between consecutive frames in milliseconds, the default value is 200 ms. Here we have chosen a delay of 20 ms. The result is shown in Fig. 3.12. When you actually observe the computer monitor rather than a figure on paper you see the damped sine wave moving toward the left. It is also possible to record the animation into an .mp4 video by inserting the line

```
ani.save('test.mp4', fps=30)
```

between lines 23 and 24. The first argument of `ani.save()` is the name of the *mp4* file, the second argument is the number of frames per second.

3.8 Exercises

Exercise 3.1 Plot $\coth(x)$ (the hyperbolic tangent of x) for x in the range $[-4, +4]$.

Exercise 3.2 Plot the function $y = 1.1 + \sin(x)$ in the range $0 \le x < 4\pi$ with a logarithmic y scale.

Exercise 3.3 Plot the function $y = sin(x)$ in the range $0 \le x < 4\pi$ with a symlog y scale. Set the linear interval to $-0.1 \le y \le 0.1$

Exercise 3.4 Write a script that draws the two-dimensional spiral of equation (in polar coordinates) $r = 10\varphi$ in the range $0 \le \varphi < 8\pi$.

Exercise 3.5 Consider a classical (non-quantum-mechanical) paramagnetic medium at temperature T. The medium consists of n permanent magnetic dipoles per unit volume, each of magnitude m_0. The magnetization M of the medium depends on the local field B_ℓ according to the law

$$M = nm_0 L(x)\,, \quad \text{where} \quad L = \coth(x) - \frac{1}{x} \quad \text{is the Langevin function, and} \quad x = \frac{m_0 B_l}{kT}\,,$$

k being the Boltzmann constant. Plot the Langevin function $L(x)$ in the range $-6 \leq x \leq 6$. Hint: both $\coth(x)$ and $1/x$ have singularities at $x = 0$, however, for small values of $|x|$, the Langevin function can be approximated by a truncation of its Taylor series:

$$L(x) = \frac{1}{3}x - \frac{1}{45}x^3 + \frac{2}{945}x^5 - \cdots$$

Chapter 4
Numerical Solution of Equations

4.1 Introduction

Solving equations, and systems of equations, of all kinds, both algebraic and transcendental, is a very frequent task in a physicist's life. Very often equations, particularly algebraic nonlinear equations and transcendental equations, have no analytical solutions. In this case we look for *approximate numerical solutions*. Some equations do not even admit solutions at all, but, of course, we shall not deal this case here! However, when tackling equations we must consider the unlucky possibility that its solution simply does not exist. In this chapter we shall consider the numerical solution of equations and systems of equations not involving the derivatives of the unknowns. Ordinary differential equations are left to Chap. 5.

4.2 Systems of Linear Equations

In this section we consider the simplest case: the solution of a system of linear equations. A linear equation is an algebraic equation in which each term is either a constant or the product of a constant and the first power of a single variable. A system of linear equations is a collection of two or more linear equations involving the same set of unknowns. The word *system* indicates that the equations are to be considered collectively, rather than individually. As an example, consider the following system of three equations in the three unknowns x_1, x_2, and x_3

$$\begin{cases} 3x_1 & -2x_2 & -x_3 & = 2 \\ 2x_1 & -2x_2 & +4x_3 & = 0 \\ -x_1 & +0.5x_2 & -1.5x_3 & = -1 \end{cases} \tag{4.1}$$

Electronic supplementary material The online version of this chapter (https://doi.org/10.1007/978-3-030-45027-4_4) contains supplementary material, which is available to authorized users.

G. Moruzzi, *Essential Python for the Physicist*,
https://doi.org/10.1007/978-3-030-45027-4_4

which can be written, in matrix form,

$$Ax = b\,, \quad \text{where} \quad A = \begin{pmatrix} 3 & -2 & -1 \\ 2 & -2 & 4 \\ -1 & 0.5 & -1.5 \end{pmatrix}, \quad x = \begin{pmatrix} x_1 \\ x_2 \\ x_3 \end{pmatrix}, \quad \text{and} \quad b = \begin{pmatrix} 2 \\ 0 \\ -1 \end{pmatrix}. \tag{4.2}$$

If we want to solve the system "by hand", we can use one of several methods, for instance *elimination of variables*, *row reduction*, *Cramer's rule*, ...

Python provides the function `numpy.linalg.solve()`, which computes the "exact" solution, $\boldsymbol{x}$, of the linear matrix equation $A\boldsymbol{x} = \boldsymbol{b}$, provided that the determinant of A is different from zero. Listing 4.1 shows how it works

Listing 4.1 LinearSyst.py

```
1 #!/usr/bin/env python3
2 import numpy as np
3 #
4 A=np.array([[3.0,-2.0,-1.0],[2.0,-2.0,4.0],[-1.0,0.5,-1.5]])
5 b=np.array([2.0,0.0,-1.0])
6 x=np.linalg.solve(A,b)
7 print('x =',x)
8 bb=np.dot(A,x)
9 print('bb =',bb)
```

Line 4 creates matrix `A` as an *array of arrays*: the single subarrays of the argument of the function `array()` are the lines of the matrix A of (4.2). Line 5 stores the vector of the constant terms into the array `b`. Finally, Line 6 solves the equations system by calling `solve()` and stores the solution into the array `x`. Line 7 prints the solution. Line 8 multiplies the matrix `A` by the vector `x` using the function `dot()`, discussed in Sect. 1.17.3, and stores the result into the array `bb`. Line 9 prints `bb`, which, if the solution is correct, must equal the array `b`. This is what you see when you run the script

```
$>LinearSyst.py
x = [ 2.00000000e+00 2.00000000e+00 5.55111512e-17]
bb = [ 2.00000000e+00 2.22044605e-16 -1.00000000e+00]
```

Thus $x_1 = 2, x_2 = 2, x_3 = 0$ is the solution of the system (4.1). Unavoidable rounding errors in numerical methods lead to 5.5×10^{-17} instead of 0 in the solution, and to 2.2×10^{-16} in the check.

4.3 Nonlinear Equations

4.3.1 Common Methods for the Numerical Solution

We are used to see most equations having both a right-hand side and a left-hand side, but here it is convenient to move all nonzero terms to the left of the equal sign, leaving only a zero at the right-hand side. In other words, we consider equations of the form

$$f(x) = 0 , \tag{4.3}$$

and search for their solutions, or sets of solutions. For nonlinear problems a root finding algorithm proceeds by iteration, starting from some approximate trial solution and improving it until some convergence criterion is satisfied. Success strongly depends on having a good first guess for the solution, and this guess usually relies on a detailed analysis of the problem. Whenever possible one should "bracket" the solution, i.e., determine an interval containing the solution. A serious problem can be posed by the existence of multiple solutions, especially if they are close to one another, and/or if they are in even number. If $f(x)$ is a continuous function, and a and b are two values such that $a < b$ and $f(a)$ and $f(b)$ have opposite signs, then (4.3) has at least a solution x_1 such that $a < x_1 < b$. But, of course, there might also be any odd number of solutions in the same interval. For a more thorough discussion of the problem see Chap. 9 of Ref. [1].

The most common methods for the numerical solution of (4.3) are the *bisection method*, the *secant Method*, the *Newton-Raphson Metod* and the *Brent Method*. Here we shall only discuss

1. the bisection method, which relies on the knowledge of the interval where the solution is located, and is very straightforward.
2. The secant method, which also requires that two points are provided, but does not require that the solution is between the two points. Providing such two points can be interpreted as providing a start point and an initial search step.

A large number of further methods are coded into functions contained in various Python packages, but we cannot discuss all of them here.

4.3.2 Bisection Method

If we have an interval (a, b) such that $a < b$ and $f(a)f(b) < 0$, the *bisection method* cannot fail, obviously provided that $f(x)$ is continuous in the interval! This is how it works:

1. evaluate the function $f(c)$ at the midpoint $c = (a + b)/2$;
2. if $f(c) = 0$ the problem is solved, but the probability for this to happen is obviously zero;
3. if $f(c)$ has the same sign as $f(a)$ replace a by c as endpoint of the interval, otherwise replace b by c. The new interval still brackets the solution, but its length is one half of the length of the original interval.
4. Go back to point 1.

At each iteration the interval containing the solution is halved, and we stop when $|b - a| < \varepsilon$, with ε the required accuracy for the solution.

A simple example of an equation which cannot be solved analytically is the following

$$5 + 4x = e^x \ , \quad \text{which we can rewrite as} \quad 5 + 4x - e^x = 0 \ . \tag{4.4}$$

We can use the code of Listing 4.2 in order to check if (4.4) has a solution in the range $0 < x < 10$

Listing 4.2 bisection01.py

```
1  #!/usr/bin/env python3
2  from numpy import exp
3  #
4  def fun(x):
5      return 5.0+4.0*x-exp(x)
6  #
```

Lines 4–5 define the function $f(x) = 5 + 4x - e^x$, whose roots we are searching.

```
6  #
7  a=0.0
8  b=10.0
9  eps=1.0e-15
10 #
```

Variables `a` and `b` are the endpoints of the investigated interval, and `eps` is the required accuracy on the solution.

```
11 fa=fun(a)
12 fb=fun(b)
13 #
14 if fa*fb>0:
15     print("wrong interval!!!",fa,fb)
16     exit()
17 #
```

Variables `fa` and `fb` are the function values at the endpoints of the interval. If the product `fa*fb` is greater than zero, `fa` and `fb` have the same sign, implying that the function crosses the x axis an even number of times (including zero!) in the interval, and the interval is rejected at lines 14–16. Function `exit()` is a standard Python function that causes the script to exit (terminate).

```
18 iter=1
19 while (b-a)>eps:
20     c=(a+b)/2.0
21     fc=fun(c)
22     if fc==0:
23         print("x = ",c)
24         exit()
25     if fc*fa>0:
26         a=c
27         fa=fc
28     else:
29         b=c
```

```
30         fb=fc
31     iter+=1
32 #
```

Lines 19–31 are the bisection loop. The midpoint `c` of the interval is evaluated at line 20, and the function value `fc` is evaluated at line 21. At lines 25–30 one of the endpoints of the interval is replaced by the midpoint, so that the interval is halved. At each iteration the counter `iter` is incremented at line 31. The loop stops when the interval width is smaller than `eps`.

```
32 #
33 print("x = ",c)
34 print("accuracy = ",'{:.2e}'.format(b-a))
35 print("f(",c,") =",fun(c))
36 print(iter," iterations needed")
```

Line 33 prints the solution, line 34 the accuracy on the solution, line 35 the function value at the solution, and line 36 the number of iterations.

This is what you observe on the terminal when you run the program

```
$> bisection01.py
x =  2.7800807820516997
accuracy =  4.44e-16
f(2.7800807820516997) = -3.5527136788e-15
55  iterations needed
```

thus, $x = 2.7800807820517000(4)$ is a solution of (4.4). If you widen the search interval at its left, by changing its endpoints at lines 7 and 8 to `a=-10.0` and `b=10.0` you get

```
$> bisection01.py
wrong interval!!! -35.0000453999 -21981.4657948
```

showing that both $f(a)$ and $f(b)$ are negative. Thus there is an even number of roots in the interval. Since we have already found one root in the right half of the interval $(-10, 10)$, the number of roots cannot be zero, and the interval must contain at least one further root. If we now change the interval endpoints at lines 7 and 8 to the values `a=-10.0` and `b=0.0` we get

```
$> bisection01.py
x =  -1.172610558265084
accuracy =  6.66e-16
f(-1.172610558265084) = 1.49880108324e-15
55  iterations needed
```

thus, also $x = -1.172610558265084(6)$ is a solution of (4.4). As we have seen, the fact that the (continuous) function has opposite signs at the interval endpoints implies that there is at least one root in the interval, but, in principle, there might be any odd number of roots. Conversely, if $f(a)$ and $f(b)$ have the same sign, this does not necessarily mean that the interval contains no root, there might be an even number of roots. Whenever possible, a rough plot of a function can be of great help in localizing the intervals where its roots are located. Then, the bisection method determines the roots with the required accuracy. Fig. 4.1 shows a plot of our $f(x)$.

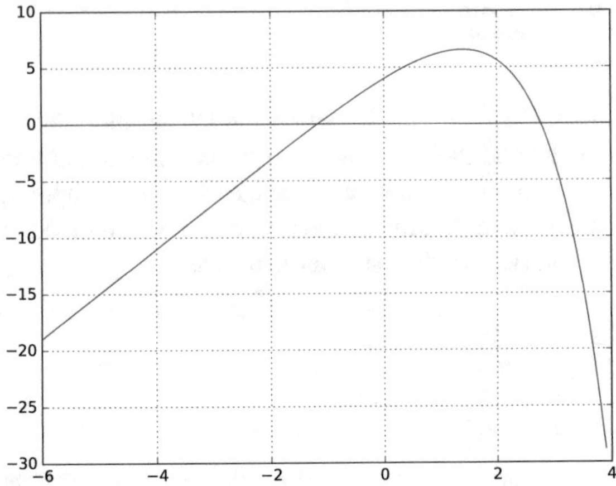

Fig. 4.1 The plot of function (4.4) between $x = -6$ and $x = 4$

4.3.3 *The Secant Method*

Also the *secant method* requires two start points, let us denote them by x_0 and x_1, possibly both close to the solution x^*. However, this method does not require that the solution lies in the (x_0, x_1) interval. Suppose that we want to find the root of the function $f(x)$ plotted in Fig. 4.2. We start from the two values x_0 and x_1 shown in the figure, and evaluate the values $y_0 = f(x_0)$ and $y_1 = f(x_1)$. Then we draw the straight line passing through the points (x_0, y_0) and (x_1, y_1), which will cross the x axis at the point

$$x_2 = x_1 - y_1 \frac{x_1 - x_0}{y_1 - y_0} , \tag{4.5}$$

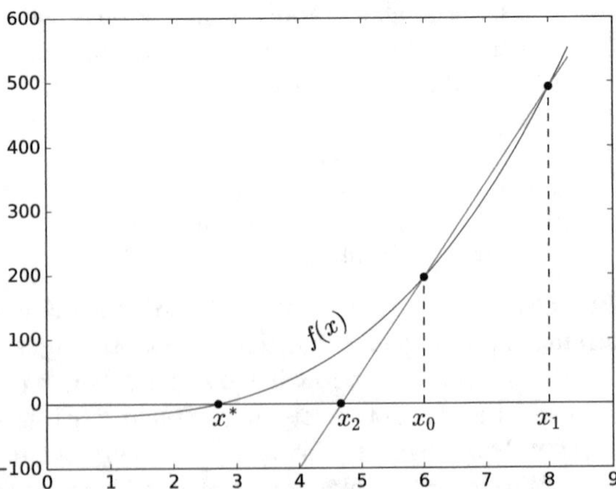

Fig. 4.2 First iteration of the secant method

which is closer to the solution x^* than both x_0 and x_1, and evaluate $y_2 = f(x_2)$. Then we proceed by iteration with

$$x_{i+1} = x_i - y_i \frac{x_i - x_{i-1}}{y_i - y_{i-1}} \,. \tag{4.6}$$

The *Newton-Raphson* method is analogous to the secant method, but it requires the first derivative of the function $f(x)$. If we do not have good starting points, the secant method and the Newton-Raphson method can fail, while the bisection method always succeeds if the function has opposite signs at the ends of the initial interval.

4.4 Root Finding with the *scipy.optimize* Package

If one has a single-unknown equation, the subpackage *scipy.optimize* provides four root finding functions, based on different algorithms. Each of these algorithms requires the endpoints of an interval in which a root is expected. In general the function `brentq()` is the best choice, but the other three functions may be useful in certain circumstances, or for academic purposes. Listing 4.3 shows a program using `scipy.optimize.bisect()` for finding the positive root of Equation (4.4)

Listing 4.3 optbisect.py

```
#!/usr/bin/env python3
from numpy import exp
from scipy.optimize import bisect
#
def fun(x):
    return 5.0+4.0*x-exp(x)
#
a=0.0
b=10.0
eps=1.0e-15
#
fa=fun(a)
fb=fun(b)
#
if fa*fb>0:
    print("wrong interval!!!",fa,fb)
    exit()
#
x=bisect(fun,a,b,xtol=eps)
print(x)
```

The three mandatory arguments of `bisect()` are the pointer to the function and the endpoints of the interval containing the root, an optional argument is `xtol`, the required absolute tolerance.

```
$> optbisect.py
2.780080782051697
```

Listing 4.4 uses `scipy.optimize.brentq()` leading, obviously, to the same result.

Listing 4.4 optbrent.py

```python
#!/usr/bin/env python3
from numpy import exp
from scipy.optimize import brentq
#
def fun(x):
   return 5.0+4.0*x-exp(x)
#
a=0.0
b=10.0
eps=1.0e-15
#
fa=fun(a)
fb=fun(b)
#
if fa*fb>0:
   print("wrong interval!!!",fa,fb)
   exit()
#
x=brentq(fun,a,b,xtol=eps)
print(x)
```

The *scipy.optimize* subpackage provides also root-finding algorithms that do not require the end points of an interval containing the solution, but only a good starting point. An example is the function `fsolve()`, used in the following listing.

Listing 4.5 fsolve_demo.py

```python
#!/usr/bin/env python3
from numpy import exp
from scipy.optimize import fsolve
#
def fun(x):
   return 5.0+4.0*x-exp(x)
#
xstart=1.0
#
x=fsolve(fun,xstart)
print(x)
```

Here we are finding a root of the same function of Listings 4.3 and 4.4, using $x = 3.0$ as starting value, obtaining the same result. The other root, $x = -1.17261056$, can be obtained by using a different starting point, for instance $x = 1.8$.

4.5 Algebraic Equations

An algebraic equation is an equation of the form $P = 0$, where P is a polynomial involving non-negative integer powers of one unknown, with coefficients usually in the field of the real numbers. The largest exponent of the unknown is the *degree of the equation*. An algebraic equation of degree n has n roots, each root being counted the number of times equal to its multiplicity. You can find the roots of an algebraic equation by the function `solveset()` of the `sympy` package, a Python library for symbolic mathematics. In order to solve the equation

$$3x^3 - 24x^2 + 57x - 36 = 0 \tag{4.7}$$

you must think of the equation in the form $f(x) = 0$, then type the following commands

```
1 >>> from sympy import solveset
2 >>> from sympy.abc import x
3 >>> solveset(3*x**3-24*x**2+57*x-36,x)
4 {1, 3, 4}
```

Line 1 imports `solveset()`, Line 2 imports x, the symbol used for the unknown, from the set of predefined symbols from `sympy.abc`. Line 3 calls `solveset()`, with $f(x)$ and x (the symbol used for the unknown) as arguments. The roots are $x_1 = 1$, $x_2 = 3$, and $x_3 = 4$.

The roots do not need to be all real, for equations of degree $n \geq 2$ there cam be pairs of complex-conjugate roots. For instance, consider the equation

$$2x^2 + 1 = 0 , \tag{4.8}$$

the roots are

```
>>> solveset(2*x**2+1,x)
{-sqrt(2)*I/2, sqrt(2)*I/2}
```

i.e., $x_{1,2} = \mp \mathrm{i}\sqrt{2}/2$. Note that `sympy` uses the symbol `I` (upper case I) for the imaginary unit i.

4.6 Systems of Nonlinear Equations

4.6.1 Equations Involving only Unknowns

When we have a system of nonlinear equations, it is usually difficult to provide a-priori intervals that contain the solutions for each unknown. Thus it is advisable to use a function like `fsolve()`, which we met in Listing 4.5 of Sect. 4.4, and whose first argument can actually be a vector of functions rather than a single function. For instance, suppose that we want to solve the system

$$\begin{cases} f_1(x_1, x_2) = 0 \\ f_2(x_1, x_2) = 0 \end{cases} \quad \text{where} \quad \begin{cases} f_1(x_1, x_2) = 4\,x_1 + 2\,x_2^2 - 4 \\ f_2(x_1, x_2) = \mathrm{e}^{x_1} + 3\,x_1 x_2 - 5\,x_2^3 + 3\,, \end{cases} \tag{4.9}$$

As start values we assume $x_1^{(0)} = 1$ and $x_2^{(0)} = 1$. The script follows

Listing 4.6 SystemSolve2.py

```
1  #!/usr/bin/env python3
2  from scipy.optimize import fsolve
3  from numpy import exp
4  #
5  def func(xvect):
6      x1,x2=xvect
7      r1=4*x1+2*x2**2-4
8      r2=exp(x1)+3*x1*x2-5*pow(x2,3)+3
9      return[r1,r2]
```

Lines 5–9 define the function `func()`, which is actually the list of the two functions $f_1(x_1, x_2)$ and $f_2(x_1, x_2)$ of (4.9). The argument `xvect` is actually the list of the two values x_1 and x_2 at which the functions are to be evaluated. Line 6 unpacks `xvect`, while lines 7 and 8 evaluate $f_1(x_1, x_2)$ and $f_2(x_1, x_2)$, respectively, Line 9 returns the list of the two function values.

```
10 xstart=(1,1)
11 sol=fsolve(func,xstart)
12 print("Solution:",sol)
13 #
14 print("Check:",func(sol))
```

Line 10 stores our start values $x_1^{(0)}$ and $x_2^{(0)}$ into `xstart`, line 11 calls `fsolve()` and stores the evaluated solutions for x_1 and x_2 into the list `sol`. Line 12 prints the solutions for x_1 and x_2 and, finally Line 14 prints the corresponding values of $f_1(x_1, x_2)$ and $f_2(x_1, x_2)$. The values of the two functions are expected to be very close to zero. This is what we get when we call `SystemSolve2.py`:

```
$> SystemSolve2.py
Solution: [ 0.43880303   1.05943095]
Check: [6.9988459472369868e-12, -1.4656365010523587e-10]
```

Thus $x_1 = 0.43880303$ and $x_2 = 1.05943095$ are two numerical solutions of the system (4.9). As always with numerical methods, the corresponding values of the two functions are not exactly zero, but very small, namely $f_1(x_1, x_2) \simeq 7.0 \times 10^{-12}$ and $f_2(x_1, x_2) \simeq -1.5 \times 10^{-10}$.

If you give wrong start values, `fsolve()` may be unable to converge to a solution. In this case you get the following warning: *the iteration is not making good progress, as measured by the improvement from the last ten iterations.*

4.6.2 Equations Involving Unknowns and Parameters

It is often convenient to pass parameter values to a function. For instance, our system (4.9) could be rewritten as

$$\begin{cases} f_1(x_1,x_2)=0 \\ f_2(x_1,x_2)=0 \end{cases} \quad \text{where} \quad \begin{cases} f_1(x_1,x_2)=a_{11}\,x_1+a_{12}\,x_2^2-c_1 \\ f_2(x_1,x_2)=a_{21}\mathrm{e}^{x_1}+a_{22}\,x_1x_2+a_{23}\,x_2^3-c_2\,, \end{cases} \tag{4.10}$$

so that the parameters a_{ij} and c_i can be changed in successive calls to the functions, related to different problems. This is how to do it:

Listing 4.7 SystemSolveParam.py

```
1  #!/usr/bin/env python3
2  from scipy.optimize import fsolve
3  from numpy import exp
4  #
5  def func(xvect,params):
6      x1,x2=xvect
7      a11,a12,a21,a22,a23,c1,c2=params
8      r1=a11*x1+a12*x2**2-c1
9      r2=a21*exp(x1)+a22*x1*x2+a23*pow(x2,3)-c2
10     return[r1,r2]
```

The function `func()` now has two arguments: the list of *unknowns*, `xvect`, and the list of *parameters*, `param`. The function `fsolve()` will solve only for `xvect`, and leave the list `param` unchanged. The variable and parameter lists are unpacked at Lines 6 and 7, respectively. Lines 8 and 9 evaluate the two functions.

```
11 a11=4
12 a12=2
13 a21=1
14 a22=3
15 a23=-5
16 c1=4
17 c2=-3
18 parlist=[a11,a12,a21,a22,a23,c1,c2]
19 xstart=(1,1)
20 sol=fsolve(func,xstart,parlist)
21 print(sol)
```

Lines 11–18 define the parameter values and store them into the list `parlist`. Line 20 passes also the argument `parlist` to `fsolve()`, which now has three arguments, the third being the list of parameters appearing in the functions. This is the result:

```
$> SystemSolveParam.py
[0.43880303 1.05943095]
```

4.7 Integration of Functions

4.7.1 Introduction

The integrals of functions, even of elementary functions, can be computed analytically only in few special cases. In most cases the only choice is numerical integration, also known as *quadrature*. Obviously, this leaves out the evaluation of indefinite integrals. Here we shall consider methods for the numerical evaluation of an integral of the form

$$I = \int_a^b f(x)\,dx \tag{4.11}$$

based on adding up the values of $f(x)$ at a sequence of x values within the integration range $[a, b]$.

A rapid survey of some common methods follows.

4.7.2 Rectangular and Trapezoidal Rules

A large class of quadrature rules are based on the division of the integration interval $[a, b]$ into some number n of smaller subintervals, and approximating $f(x)$ by a polynomial of low degree (which is easy to integrate) in each subinterval. For simplicity we shall assume subintervals of equal length

$$\Delta x = \frac{b-a}{n} , \tag{4.12}$$

implying the following endpoints for the i-th subinterval

$$a_i = a + i\,\Delta x\ , \quad b_i = a + (i+1)\,\Delta x\ , \quad \text{with} \quad b_i = a_{i+1}\ , \quad \text{and} \quad i = 0, 1, 12, \ldots, n-1\ . \tag{4.13}$$

The simplest method is to let the approximate function be a constant function (thus, a polynomial of degree zero) within each subinterval, passing through the point

$$\Big(x_i, f(x_i)\Big)\ , \quad \text{with} \quad x_i = a + (2i+1)\,\frac{\Delta x}{2} = a_i + \frac{\Delta x}{2}\ . \tag{4.14}$$

This is called the midpoint rule or rectangle rule. The resulting approximation is

$$I = \int_a^b f(x)\,dx \simeq \Delta x \sum_{i=0}^{n-1} f(x_i)\ , \tag{4.15}$$

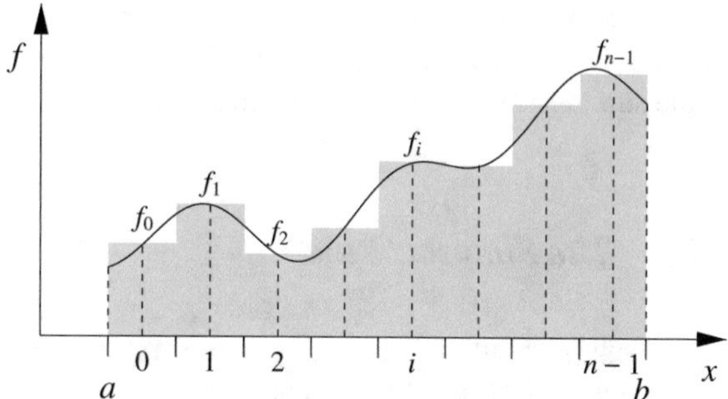

Fig. 4.3 Integration by the rectangle rule

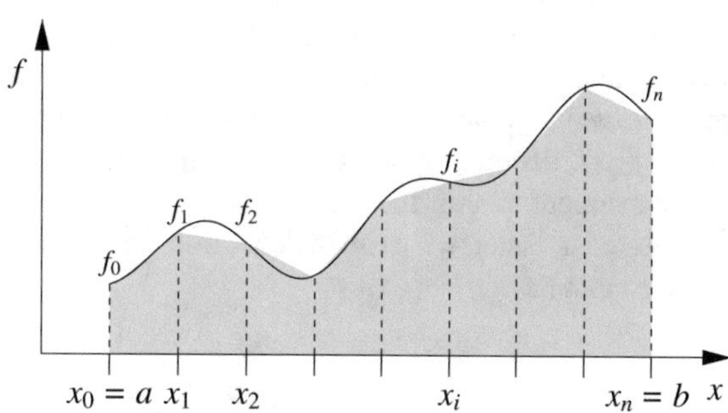

Fig. 4.4 Integration by the trapezoidal rule

which corresponds to approximate the area under the figure curve by the shaded area of Fig. 4.3. The next step is approximating the function by a polynomial of degree one (a straight line) in each subinterval. For this, it is convenient to label $x_0 = a$, $x_n = b$, and $x_i = a_i = b_{i-1}$ for $1 \leqslant i \leqslant n-1$, as in Fig. 4.4. The i-th subinterval has x_i and x_{i+1} as upper and lower limits, respectively, and the corresponding approximating straight line passes through the points $\big(x_i, f(x_i)\big)$ and $\big(x_{i+1},\ f(x_{i+1})\big)$, so that the area below the function curve is approximated by a sum of trapezoids (shaded area in Fig. 4.4). The area of the i-th trapezoid is

$$\frac{f(x_i) + f(x_{i+1})}{2}\,\Delta x\ , \tag{4.16}$$

and the approximation for the integral is

$$\begin{aligned} I &= \int_a^b f(x)\,dx \\ &\simeq \Delta x \left[\frac{f(x_0) + f(x_n)}{2} + \sum_{i=1}^{n-1} f(x_i) \right] . \end{aligned} \tag{4.17}$$

Note that n function values $f_0 \dots f_{n-1}$, corresponding to the subinterval midpoints of Fig. 4.3, are involved in (4.15), while $n+1$ function values $f_0 \dots f_n$, corresponding to the subinterval upper and lower endpoints of Fig. 4.4, are involved in (4.17).

4.7.3 The Simpson Rule

In the next (and last, at least in this context) step we approximate the area of a couple of adjacent subintervals by the area below a polynomial of degree two, i.e., a parabolic arc. This is shown for the couple of subintervals between x_{i-1} and x_{i+1} of Fig. 4.5. Thus, the whole integration interval $[a, b]$ must be divided into an even number n of subintervals, and we must have $i = 2m + 1$, with $m = 0, \dots, (n/2) - 1$. The approximating parabola must pass through the three points (x_{i-1}, f_{i-1}), (x_i, f_i) and (x_{i+1}, f_{i+1}), denoted by the black dots in Fig. 4.5. For each couple of subintervals it is convenient to shift the x origin so that $x_{i-1} = -\Delta x$, $x_i = 0$ and $x_{i+1} = \Delta x$. In this reference frame we write the equation for the parabola as $y = \alpha_i x^2 + \beta_i x + \gamma_i$, and we must have

$$\begin{aligned} \alpha_i\,(\Delta x)^2 - \beta_i\,\Delta x + \gamma_i &= f_{i-1}\,, \\ \gamma_i &= f_i\,, \\ \alpha_i\,(\Delta x)^2 + \beta_i\,\Delta x + \gamma_i &= f_{i+1}\,. \end{aligned} \tag{4.18}$$

Solving for α_i, β_i and γ_i we obtain

$$\alpha_i = \frac{f_{i-1} - 2f_i + f_{i+1}}{2\,(\Delta x)^2}\,, \quad \beta_i = \frac{f_{i+1} - f_{i-1}}{2\Delta x}\,, \quad \gamma_i = f_i\,, \tag{4.19}$$

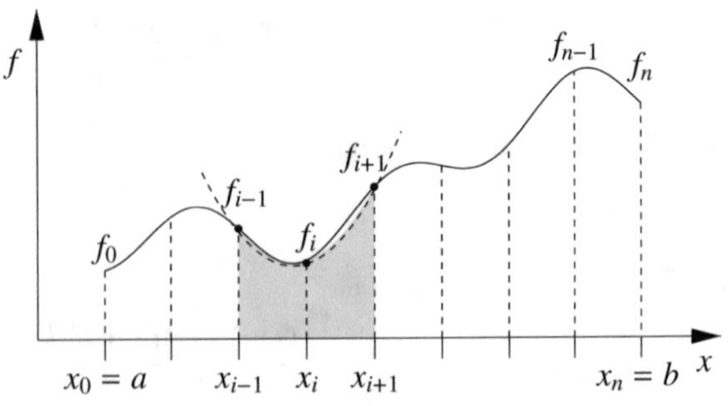

Fig. 4.5 Integration by the Simpson rule

and the area below the parabolic arc is

$$\Delta S_i = \int_{-\Delta x}^{\Delta x} \left(\alpha_i x^2 + \beta_i x + \gamma_i\right) \mathrm{D}x = \left[\alpha_i \frac{x^3}{3} + \beta_i \frac{x^2}{2} + \gamma_i x\right]_{-\Delta x}^{\Delta x} = \left(\frac{1}{3} f_{i-1} + \frac{4}{3} f_i + \frac{1}{3} f_{i+1}\right) \Delta x \,. \tag{4.20}$$

The approximate integral is thus

$$I \simeq S = \sum_i \Delta S_i = \frac{\Delta x}{3} \left(f_0 + 4 f_1 + 2 f_2 + 4 f_3 + 2 f_4 + \cdots + 4 f_{n-1} + f_n\right) \,. \tag{4.21}$$

4.7.4 *The* scipy.integrate.simps *Function*

The *scipy.integrate* subpackage provides the `simps()` function discussed in Appendix C.1, which integrates the samplings of a function using the Simpson rule. As a simple example, Listing 4.8 evaluates the integral of the Gaussian function e^{-x^2} between -5 and 5 using `simps()`.

$$\int_{-5}^{5} e^{-x^2} dx \tag{4.22}$$

Listing 4.8 CheckSimps.py

```
1 #!/usr/bin/env python3
2 import numpy as np
3 from scipy.integrate import simps
4 #
5 x=np.linspace(-5.0,5.0,31)
6 y=np.exp(-x**2)
7 integ=simps(y,x,even='avg')
8 print(integ)
9 print(np.sqrt(np.pi))
```

Line 5 generates an array `x` comprising 31 equally spaced numbers between -5 and 5. Line 6 generates an array `y` such that $y_i = e^{-x_i^2}$ for $i = 0, 1, 2, \ldots, 30$. Line 7 evaluates the variable `integ` which equals the integral (4.22) as evaluated by the Simpson rule. The mandatory arguments of `simps()` are the array `y`, containing the sampling of the function to be integrated, and the array `x`, containing the abscissas of the sampling points. The optional argument `even` is active only if the number of sampling points n is even, thus corresponding to an odd number $n - 1$ of subintervals. In this case `even` can have the following three values, corresponding to different behaviors of `simps()`:

- `even='first'` integrate the first $n - 2$ intervals with the Simpson rule, and the last interval with the trapezoidal rule;

- `even='last'` integrate the first interval with the trapezoidal rule and the last $n-2$ intervals with the Simpson rule;
- `even='avg'` average the results of the above two methods;

Lines 8 and 9 print `integ` and $\sqrt{\pi}$ for a comparison, since we know that

$$\int_{-\infty}^{\infty} e^{-x^2} dx = \sqrt{\pi} \; . \tag{4.23}$$

This is what we obtain when we run the script

```
$>CheckSimps.py
1.77245385117
1.77245385091
```

4.8 Exercises

Exercise 4.1 Solve the following system of linear equations

$$\begin{cases} x_1 \; +3x_2 \; -2x_3 = 5 \\ 3x_1 \; +5x_2 \; +6x_3 = 7 \\ 2x_1 \; +4x_2 \; +3x_3 = 8 \end{cases}$$

Exercise 4.2 Solve the transcendental equation

$$\cos(2\sqrt{x}) + 1.5 = \sqrt{4-x} \; .$$

Exercise 4.3 Evaluate the four roots of the algebraic equation

$$x^4 - 8x^3 + 17x^2 + 22x - 104 = 0 \; .$$

Exercise 4.4 Solve the following system of transcendental equations

$$\begin{cases} 2x^2 - \sin(x) + xy - 3 = 0 \\ 3x + y^2 + \cos(xy) - 5 = 0 \end{cases}$$

Exercise 4.5 Evaluate the integral

$$\int_{-2}^{2} \frac{dx}{1+x^2}$$

using steps $\Delta x = 0.2$, $\Delta x = 0.1$, and $\Delta x = 0.05$.

Reference

1. W.H. Press, S.A. Teukolsky, W.T. Vetterling, B.P. Flannery, *Numerical Recipes in C*, 2nd edn. (Cambridge University Press, Cambridge, New York, Victoria, 1992)

Chapter 5
Numerical Solution of Ordinary Differential Equations (ODE)

5.1 Introduction

A differential equation is an equation that contains an unknown function (rather than an unknown value) to be determined, and at least one of its derivatives with respect to an independent variable. A differential equation is called linear if it is a linear polynomial in the unknown function and its derivatives. Here we shall be concerned only with linear differential equations. If the unknown function depends only on a single independent variable, the differential equation can contain only ordinary derivatives, and it is called an *ordinary differential equation*. On the other hand, if the unknown function depends on several independent variables, and the equation involves partial derivatives of the unknown function instead of ordinary derivatives, the equation is called a *partial differential equation*. In this chapter we are concerned with ordinary differential equations.

A problem involving ordinary differential equations (ODEs) of any order can always be reduced to the study of a system of first-order differential equations. For example the second-order differential equation containing the unknown function $y(x)$

$$\frac{\mathrm{d}^2 y}{\mathrm{d}x^2} + f(x, y)\,\frac{\mathrm{d}y}{\mathrm{d}x} = g(x, y) \tag{5.1}$$

can be rewritten as the system of two first-order differential equations

$$\begin{aligned} \frac{\mathrm{d}y}{\mathrm{d}x} &= z(x, y) \\ \frac{\mathrm{d}z}{\mathrm{d}x} &= g(x, y) - f(x, y)\,z(x, y), \end{aligned} \tag{5.2}$$

Electronic supplementary material The online version of this chapter (https://doi.org/10.1007/978-3-030-45027-4_5) contains supplementary material, which is available to authorized users.

G. Moruzzi, *Essential Python for the Physicist*,
https://doi.org/10.1007/978-3-030-45027-4_5

where z is a new unknown function. This is the standard procedure at the basis of the methods for the numerical solution of ODEs. Thus, the generic problem involving ordinary differential equations can be reduced to the study of a system of N coupled first-order differential equations involving N unknown functions y_i, with $i = 1, 2, \ldots, N$, of the form

$$\frac{\mathrm{d}y_i}{\mathrm{d}x} = f_i(x, y_1, \ldots, y_N), \tag{5.3}$$

where the functions $f_i(x, y_1, \ldots, y_N)$ are known. We know that the solutions of differential equations are not completely specified unless appropriate boundary conditions on the y_i are given.

Boundary conditions are divided into two broad categories:

- Initial value problems, where all the y_i are known at some starting value x_0 of the independent variable. In this case it is desired to find the y_i at some final point x_f, or at some discrete list of points (for example, at tabulated intervals).
- In two-point boundary value problems, on the other hand, boundary conditions are specified at more than one x value. Typically, some of the conditions will be specified at the starting point x_0, and the remaining conditions at the final point x_f.

In the following we shall consider the initial value problem up to Sect. 5.4, while we shall have a look at the (in general more difficult) two-point boundary value problems in Sect. 5.5.

Most routines for the numerical solution of differential equations are based on the replacement of the differentials $\mathrm{d}y_i$ and $\mathrm{d}x$ appearing in the equations by small, but *finite*, steps Δy_i and Δx: they are thus called *finite-difference methods*. After the replacement, the equations are multiplied by Δx, thus leading to algebraic first-order formulas for the change in the functions Δy_i when the independent variable x is increased by one *step* Δx. At the limit of very small stepsizes a good approximation of the differential equation is obtained. The simplest implementation of this procedure is Euler's method, which is conceptually very important, but not recommended for practical use. However, all more refined methods rely on Euler's method as a starting point.

5.2 Euler and Runge–Kutta Methods

If we wish to solve numerically the differential equation

$$\frac{\mathrm{d}y}{\mathrm{d}x} = f(x, y), \tag{5.4}$$

with the initial condition $y(x_0) = y_0$, the Euler method, which is the simplest method, provides a table of values y_n and x_n, obtained by the recursive formulas

$$x_{n+1} = x_n + \Delta x, \quad y_{n+1} = y_n + f(x_n, y_n)\,\Delta x \; + O(\Delta x^2), \tag{5.5}$$

where usually the spacing Δx is kept constant. The main disadvantage of this formula is that it is unsymmetrical: it advances the solution by a step Δx, but uses derivative information only at one end (the beginning) of the step. The error per step is of the order of Δx^2.

An improvement to the Euler method is the second-order Runge-Kutta method [nth order means that the error per step is of order $O(\Delta x^{n+1})$], defined by the sequence

$$\Delta y^{(1)} = f(x_n, y_n)\,\Delta x, \tag{5.6}$$

$$\Delta y^{(2)} = f\left(x_n + \frac{1}{2}\Delta x,\; y_n + \frac{1}{2}\Delta y^{(1)}\right)\Delta x, \tag{5.7}$$

$$y_{n+1} = y_n + \Delta y^{(2)} + O\left(\Delta x^3\right), \tag{5.8}$$

where the Euler method of (5.6) is used to extrapolate the midpoint of the interval $(x_n + \Delta x/2,\; y_n + \Delta y^{(1)}/2)$ where we evaluate the derivative used for the more accurate $\Delta y^{(2)}$ value of (5.7).

But the most often used formula is the fourth-order Runge–Kutta method, which proceeds as follows

$$\Delta y^{(1)} = f(x_n, y_n)\,\Delta x, \tag{5.9}$$

$$\Delta y^{(2)} = f\left(x_n + \frac{1}{2}\Delta x,\; y_n + \frac{1}{2}\Delta y^{(1)}\right)\Delta x, \tag{5.10}$$

$$\Delta y^{(3)} = f\left(x_n + \frac{1}{2}\Delta x,\; y_n + \frac{1}{2}\Delta y^{(2)}\right)\Delta x, \tag{5.11}$$

$$\Delta y^{(4)} = f\left(x_n + \Delta x,\; y_n + \Delta y^{(2)}\right)\Delta x, \tag{5.12}$$

$$y_{n+1} = y_n + \frac{\Delta y^{(1)}}{6} + \frac{\Delta y^{(2)}}{3} + \frac{\Delta y^{(3)}}{3} + \frac{\Delta y^{(4)}}{6} + O\left(\Delta x^5\right), \tag{5.13}$$

Thus the derivative is evaluated four times at each step: once at the initial point, twice at trial midpoints, and once at a trial endpoint, and the final y increment is evaluated as a weighted average of the four $\Delta y^{(i)}$ values. Consequently, the fourth-order Runge–Kutta method is superior to the second-order method if it can use a step at least twice as large achieving at least the same accuracy. This is very often the case, but not always.

5.3 The *scipy.integrate.odeint* Function

The `scipy.integrate.odeint()` function integrates a system of ordinary differential equations of the type (5.3) using *lsoda* from the FORTRAN library *odepack*. It solves the initial value problem for stiff or non-stiff systems of first order ODE's. We shall use it in the form

```
y=odeint(f,y0,x,args=(params,))
```

where `f(y,x0)` evaluates the derivative of `y` at `x0`, `y0` is an array containing the initial conditions on `y`, `x` is the sequence of x points at which `y` should be evaluated. The initial value point should be the first element of this sequence. Argument `args` is a tuple containing extra arguments to pass to the function. Things are probably made clearer by the example of Sect. 5.4.

5.4 Large-Amplitude Pendulum

The equation of motion for a simple pendulum without friction is

$$\ell^2 m \frac{\mathrm{d}^2\vartheta}{\mathrm{d}t^2} = -\ell m g \sin\vartheta, \quad \text{which reduces to} \quad \frac{\mathrm{d}^2\vartheta}{\mathrm{d}t^2} = -\frac{g}{\ell}\sin\vartheta, \tag{5.14}$$

where ℓ is the length of the pendulum, m its mass, $\ell^2 m$ its moment of inertia with respect to the pivot O of Fig. 5.1, and $\ell m g \sin\vartheta$ the torque of the gravity force $m\mathbf{g}$ with respect to the pivot. Equation (5.14) has no anlytical solution, but in the case of small oscillation amplitudes one can approximate $\sin\vartheta \simeq \vartheta$, obtaining the well known equation for the classical harmonic oscillator

$$\frac{\mathrm{d}^2\vartheta}{\mathrm{d}t^2} = -\frac{g}{\ell}\vartheta, \tag{5.15}$$

as found in any introductory Physics book. Here, however, we are interested in the numerical solution of the more general Eq. (5.14). The numerical solution can be obtained by running script 5.1.

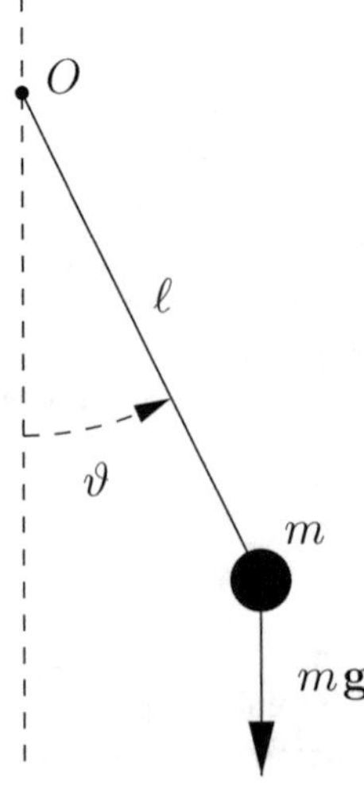

Fig. 5.1 Pendulum

Listing 5.1 PlotPendulum.py

```
1  #!/usr/bin/env python3
2  #coding: utf8
3  import numpy as np
4  from scipy.integrate import odeint
5  import matplotlib.pyplot as plt
6  #
7  def dydt(y0,t, params):
8     theta,omega = y0
9     GdivL,=params
10    derivs = [omega,-GdivL*np.sin(theta)]
11    return derivs
12 #
```

Lines 7–11 define the function `dydt()` required as first argument by `odeint()`. The arguments of `dydt()` are the array of initial conditions `y0`, the array of times `t` at which the derivatives are required, and the array of additional parameters `params` needed for evaluating the derivatives. We want to solve the second-order differential equation (5.14), which, according to (5.2), can be rewritten as the system of two first-order differential equations

$$\begin{aligned} \frac{\mathrm{d}\vartheta}{\mathrm{d}t} &= \omega \\ \frac{\mathrm{d}\omega}{\mathrm{d}t} &= -\frac{g}{l}\sin\vartheta. \end{aligned} \tag{5.16}$$

Line 8 unpacks the initial values for θ and ω from the array `y0`. Line 9 copies our only parameter, `GdivL`, corresponding to the ratio g/l, from the single-element array `params`, see Sect. 1.12.2. Line 10 builds the list `derivs`, comprising the time derivative of ϑ (simply ω), and the time derivative of ω, given by the second of (5.16). The function returns `derivs` at Line 11.

```
13 theta0=np.pi/2.0
14 omega0=0.0
15 y0=[theta0,omega0]
16 #
17 GdivL=4.9
18 params=[GdivL]
19 Omega=np.sqrt(GdivL)
20 period=2.0*np.pi/Omega
21 #
```

Lines 13–14 fix the initial conditions of the pendulum motion: the pendulum starts from a horizontal position, $\vartheta(0) = \pi/2$, with zero angular velocity, $\omega(0) = 0$. The initial conditions are stored in the array `y0` at line 15. The only parameter needed by the function `dydt()` is the ratio g/ℓ, defined at line 17 and stored into the list `params` at line 18. A pendulum length $\ell = 2$ m is assumed, with $g = 9.8$ m/s^2. Variable `Omega` at line 19 is the angular frequency for the small amplitude (harmonic) oscillations, $\Omega = \sqrt{g/\ell}$, while `period` at line 20 is the oscillation period of the purely harmonic oscillator, $T = 2\pi/\Omega$, i.e., $T \simeq 2.838$ s in our case.

```
22 t=np.linspace(0.0,period,101)
23 ThetaHar=theta0*np.cos(Omega*t)
24 #
25 sol=odeint(dydt,y0,t,args=(params,))
26 theta=sol[:,0]
27 plt.plot(t,theta,'k')
28 plt.plot(t,ThetaHar,'k--')
```

Array `t`, defined at line 22, comprises the times, in seconds, at which $\vartheta(t)$ and $\omega(t)$ are to be evaluated, from $t = 0$ s up to $t = 2.838$ s. A total of 101 equally spaced times, corresponding to 100 time intervals, are requested in the interval $0 \leqslant t \leqslant T$. Line 23 evaluates the angular position $\vartheta_{\text{harmonic}}$, at the same times, of a hypothetical pendulum following a purely harmonic motion with angular frequency Ω, for comparison to our large-amplitude case. At line 25 function `odeint` solves the system (5.16) numerically at the times specified by the list `t`, with the initial conditions specified by the list `y0`. The parameter g/ℓ is passed by the list `params` [see Appendix C.2 for the arguments of `odeint()`]. The solutions are stored into the matrix `sol`. Line 26 copies the ϑ_i values at the required instants from matrix `sol` into the vector `theta`. Column `sol(:,1)` comprises the angular velocity values ω_i at the same instants. Line 27 plots the large-amplitude pendulum positions, using a black (`'k'`) solid line, while line 28 plots the harmonic positions using a dashed black line (`'k--'`).

```
29 plt.axhline(linewidth=1, color='k')
30 plt.rcParams.update({'font.size': 18})
31 plt.grid()
32 plt.xlabel(r'$t$/s', fontsize=22)
33 plt.ylabel(r'$\vartheta$/rad',fontsize=22)
34 plt.text(2.25,1.0,'Harmonic',fontsize=16,rotation=65)
35 plt.text(2.25,0.4,'Large Amplitude',fontsize=16,rotation=65)
36 plt.tight_layout()
37 plt.savefig('LargAmpl00.pdf',format='pdf',dpi=1000)
38 plt.show()
```

Line 29 plots a horizontal black line corresponding to the $\vartheta = 0$ axis of the plot. Line 30 sets the font size for the plot. Line 31 draws a grid. Lines 32 and 33 label the horizontal and vertical axes, respectively. Line 34 writes a text into the figure at $x = 2.25$, $y = 1.0$, rotated by 65 degrees, which labels the curve for the harmonic motion. Line 35 does the same for the curve corresponding to the large-amplitude-pendulum motion, The font-size values at lines 30, 32, 33, 34 and 35 are best determined by trial and error, in order to obtain the best result for Fig. 5.2. The same is true for the rotation angles at lines 34 and 35. Line 36 forces the figure to include the axis labels. Line 37 saves the result into a pdf figure, in our case Fig. 5.2. Line 38 shows the plot on the monitor. If you wish to tabulate $\vartheta(t)$, the last lines of the code can be changed as follows

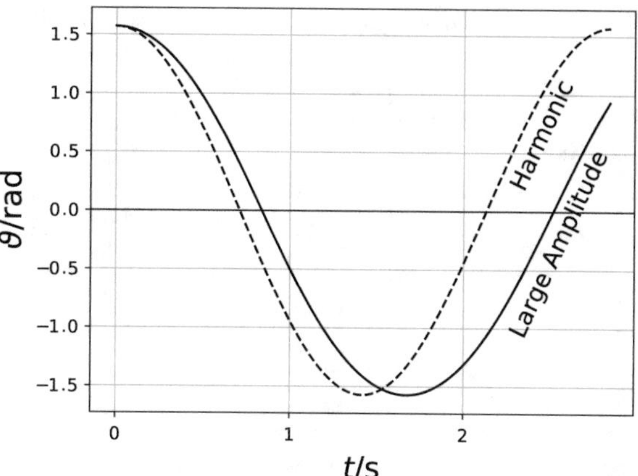

Fig. 5.2 Time dependence of the deflection angle of the large-amplitude pendulum (solid line) compared to a purely harmonic motion (dashed line) The pendulum starts from a horizontal position, $\vartheta_0 = \pi/2$, with zero initial angular velocity

```
36  plt.tight_layout()
37  plt.savefig('LargAmpl00.pdf',format='pdf',dpi=1000)
38  hnd=open("large_amplitude.txt",'w')
39  i=0
40  hnd.write('  i    sec     theta    harmonic\n\n')
41  while i<len(theta):
42    hnd.write('{:3d}{:8.3f}{:8.3f}{:8.3f}\n'.format(i,t[i],theta[i],ThetaHar[i]))
43    i+=1
44  hnd.close()
45  plt.show()
```

Line 38 creates the file `large_amplitude.txt` in `write` mode (`'w'`), and associates it to the *file handler* `hnd`. Line 38 sets the counter `i` to zero, line 39 writes the table headers into the table, and the loop 40–42 writes the single lines of the table, inserting the index `i` in column 0, the time `t[i]` in column 1, the large-amplitude angular position $\vartheta(t)$ in column 2, and the "harmonic" angular position $\vartheta_{\text{harmonic}}(t)$ in column 3. Line 43 closes the file containing the table. The first few lines of the resulting table are shown in Table 5.1.

The file `large_amplitude.txt` can now be opened and read by any editor, or opened in `read` mode by another Python script using the command

```
hnd=open('large_amplitude.txt','r')
```

then the single lines can be read by the command `hnd.readline()`.

The difference between the large-amplitude motion and the harmonic motion is much more dramatic if we let the pendulum start from an upward, almost vertical position. instead of a horizontal position. Figure 5.1 compares the large-amplitude motion starting from an angle $\vartheta_0 = 170° \simeq 2.967$ rad, instead of the $\vartheta_0 = 90° = \pi/2$ rad of Fig. 5.2. The initial angular velocity iz zero in both cases. For this we simply need to change Line 13 of Listing as follows

```
13  theta0=170.0*np.pi/180.0
```

Table 5.1 Deflection angles of the large- and small- amplitude pendulums as functions of time

i	sec	Theta	Harmonic
0	0.000	1.571	1.571
1	0.028	1.569	1.568
2	0.057	1.563	1.558
3	0.085	1.553	1.543
4	0.114	1.539	1.521
5	0.142	1.521	1.494
6	0.170	1.500	1.460
7	0.199	1.474	1.421
8	0.227	1.445	1.376
9	0.255	1.411	1.326
10	0.284	1.374	1.271

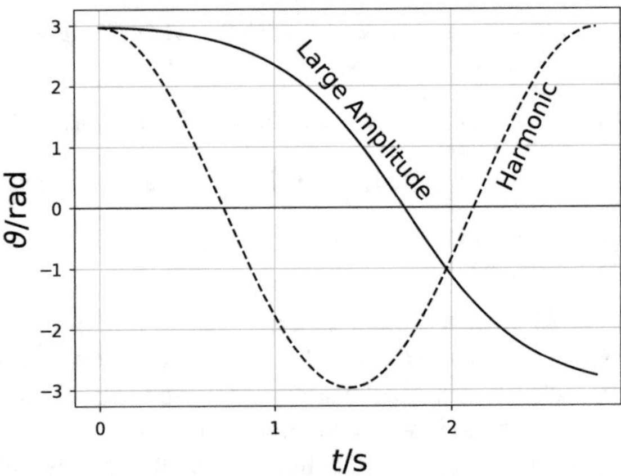

Fig. 5.3 Same as Fig. 5.2, but the pendulum starts at $\vartheta_0 \simeq 2.967$ rad (170°) with zero initial angular velocity

The result is shown in Fig. 5.3, where, again, the solid lines corresponds to the large-amplitude motion, and the dashed line to the pure harmonic motion.

5.5 The Shooting Method

The differential equation of Sect. 5.4 was second-order because Newton's second law involves a second-order derivative. We know that the general solution of an nth-order differential equation contains n arbitrary independent constants of integration, and in the case of our pendulum we arbitrarily chose two initial conditions, namely the initial position $\vartheta(0) = \pi/2$ and the initial angular velocity $\dot{\vartheta} = \omega = 0$.

We conclude this chapter considering two problems involving second-order ordinary differential equations where we are interested in solutions determined by the

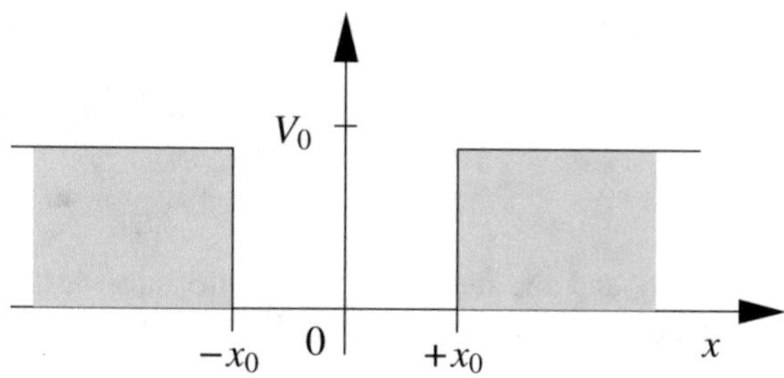

Fig. 5.4 Finite square well

conditions at two boundaries: one initial and one final condition. Both problems involve quantum mechanics and the time-independent one-dimensional Schrödinger equation, which involves the second-order derivative of the wave function with respect to the space coordinate.

5.5.1 *The Finite Square Well*

Our first example is a simple one-dimensional problem which, in spite of being simple, has no analytical solution: the *finite square well*. In this problem a particle of mass m is confined to a box of width $2x_0$ which has finite potential walls of height V_0, as shown in Fig. 5.4. Denoting the particle position by x, we can choose a coordinate system where the potential energy of the particle can be written as

$$V(x) = \begin{cases} 0 & \text{if } \ |x| \le x_0, \\ V_0 & \text{if } \ |x| > x_0. \end{cases} \tag{5.17}$$

The time-independent Schrödinger equation for the particle is thus

$$\left(V(x) - \frac{\hbar^2}{2m} \frac{\mathrm{d}^2}{\mathrm{d}x^2} \right) \psi(x) = E\psi(x), \tag{5.18}$$

which can be rewritten

$$\left(V_0 H(|x| - x_0) - \frac{\hbar^2}{2m} \frac{\mathrm{d}^2}{\mathrm{d}x^2} \right) \psi(x) = E\psi(x), \tag{5.19}$$

where E is the energy eigenvalue, and $H(x)$ is the Heaviside step function, defined as $H(x) = 0$ if $x < 0$, and $H(x) = 1$ if $x > 0$. We have a discrete spectrum of bound states for eigenvalues $E_n < V_0$, and a continuous spectrum of free states with eigenvalues $E > V_0$. Here we are interested only in bound states. Both for bound and free states, the wavefunction has three different mathematical expressions, depending on whether the particle is at the left of the box, inside the box or at the right of the box:

$$\psi(x) = \begin{cases} \psi_a(x), & \text{if } x < -x_0, \\ \psi_b(x), & \text{if } -x_0 < x < x_0, \\ \psi_c(x), & \text{if } x > x_0. \end{cases} \tag{5.20}$$

For both bound and free states the solutions inside the well, i.e., for $-x_0 < x < x_0$, where the potential energy is zero, have the form

$$\psi_b(x) = A\sin(kx) + B\cos(kx), \quad \text{with} \quad k = \frac{\sqrt{2mE}}{\hbar}, \tag{5.21}$$

where A and B are two constants to be determined. The solutions for bound states $(E_n < V_0)$ outside of the well have the form

$$\begin{aligned} \psi_a(x) &= C\,\mathrm{e}^{\alpha x}, \\ \psi_c(x) &= D\,\mathrm{e}^{-\alpha x}, \quad \text{both with} \quad \alpha = \frac{\sqrt{2\mathrm{m}(\mathrm{V}_0 - \mathrm{E})}}{\hbar}, \end{aligned} \tag{5.22}$$

where C and D are two further constants to be determined, and we are disregarding the solutions diverging for $x \to -\infty$ or $x \to \infty$. With our coordinate choice the potential energy $V(x)$ is an even function of x, therefore our eigenfunctions must be either even or odd functions of x, implying that either $A = 0$ and $C = D$ (even functions), or $B = 0$ and $C = -D$ (odd functions). Finally, a wave function $\psi(x)$ and its first derivative $\mathrm{d}\psi(x)/\mathrm{d}x$ must be continuous everywhere, including at $x = \pm x_0$, where the potential $V(x)$ is discontinuous. This implies that the logarithmic derivative

$$\frac{\mathrm{d}\ln[\psi(x)]}{\mathrm{d}x} = \frac{1}{\psi(x)}\frac{\mathrm{d}\psi(x)}{\mathrm{d}x} \tag{5.23}$$

must be continuous at $x = \pm x_0$. Therefore we must have

$$\begin{aligned} &\text{even:} \quad -\frac{kB\sin(kx_0)}{B\cos(kx_0)} = -\frac{\alpha C\,\mathrm{e}^{-\alpha x_0}}{C\,\mathrm{e}^{-\alpha x_0}} \quad \Rightarrow \quad \tan(kx_0) = \frac{\alpha}{k} \quad \Rightarrow \quad \tan\left(\frac{\sqrt{2mE}}{\hbar}x_0\right) = \sqrt{\frac{V_0 - E}{E}}, \\ &\text{odd:} \quad \frac{kB\cos(kx_0)}{B\sin(kx_0)} = -\frac{\alpha C\,\mathrm{e}^{-\alpha x_0}}{C\,\mathrm{e}^{-\alpha x_0}} \quad \Rightarrow \quad \cot(kx_0) = -\frac{\alpha}{k} \quad \Rightarrow \quad \cot\left(\frac{\sqrt{2mE}}{\hbar}x_0\right) = -\sqrt{\frac{V_0 - E}{E}}. \end{aligned} \tag{5.24}$$

The above conditions cannot be satisfied by arbitrary values of the particle energy E. Only certain energy values (the energy eigenvalues), which solve one of the (5.24) for E, are allowed. Thus the energy levels of the particle such that $E < V_0$ are discrete, and the corresponding eigenfunctions are bound states. Equation (5.24) cannot be solved analytically, and the values of E for which they hold must be found numerically. A possibility is to write them in the form $f(E) = 0$, with, for instance,

$$f(E) = \tan\left(\frac{\sqrt{2mE}}{\hbar}x_0\right) - \sqrt{\frac{V_0 - E}{E}} \tag{5.25}$$

for the even case, and then use one of the numerical root-finding algorithms discussed in Chap. 4 to determine the allowed values of E.

In the following we prefer to consider another numerical method, which will provide both the energy eigenvalues and the discretization of the eigenfunctions of the finite square well.

For this it is convenient to write (5.19) in terms of dimensionless quantities. We start by dividing (5.19) by V_0, obtaining

$$\frac{\hbar^2}{2mV_0}\frac{\mathrm{d}^2\psi_n(x)}{\mathrm{d}x^2} = \left[(H(|x| - x_0) - \frac{E_n}{V_0}\right]\psi_n(x), \tag{5.26}$$

where we are numbering the eigenfunction $\psi_n(x)$ and the energy eigenvalues E_n because we are dealing with a discrete spectrum. Then we introduce the first dimensionless variable

$$\xi = \frac{x}{\alpha}, \quad \text{such that} \quad x = \alpha\xi, \tag{5.27}$$

where $\alpha = \hbar/\sqrt{2mV_0}$ has the dimensions of a length, and obtain

$$\frac{\mathrm{d}^2\psi_n(\xi)}{\mathrm{d}\xi^2} = \left[H(|\xi| - \xi_0) - \frac{E_n}{V_0}\right]\psi_n(\xi), \quad \text{where} \quad \xi_0 = \frac{x_0}{\alpha} = \frac{\sqrt{2mV_0}}{\hbar}x_0. \tag{5.28}$$

If we introduce the further dimensionless quantities $W_n = E_n/V_0$, corresponding to the energy eigenvalues measured in units of the well depth V_0, our final equation is

$$\frac{\mathrm{d}^2\psi_n(\xi)}{\mathrm{d}\xi^2} = \left[H(|\xi| - \xi_0] - W_n\right]\psi_n(\xi). \tag{5.29}$$

As discussed in Sect. 5.1, the second-order differential equation (5.29) is equivalent to the following system of two first-order ordinary differential equations (ODE)

$$\begin{aligned}\frac{\mathrm{d}\psi_n(\xi)}{\mathrm{d}\xi} &= \chi_n \\ \frac{\mathrm{d}\chi_n(\xi)}{\mathrm{d}\xi} &= \left[H(|\xi| - \xi_0] - W_n\right]\psi_n(\xi),\end{aligned} \tag{5.30}$$

which we shall use for the numerical solution. If an eigenvalue W_n, and the corresponding initial values for $\psi_n(0)$ and $\chi_n(0)$, were known, solving the system (5.30) numerically would provide the discretization of the functions $\psi_n(\xi)$ and $\chi_n(\xi) = \mathrm{d}\psi_n(\xi)/\mathrm{d}\xi$. In the present case the potential is even, thus the eigenfunctions $\psi_n(\xi)$ are either even, with $\chi_n(0) = \mathrm{d}\psi_n(0)/\mathrm{d}\xi = 0$, or odd, with $\psi_n(0) = 0$. The eigenfunction corresponding to the ground state, $\psi_0(\xi)$, is even, and the eigenfunctions ψ_n, with $n > 0$, corresponding to the higher energy levels, are alternatively odd (n odd), and even (n even). If $\psi_n(\xi)$ is an eigenfunction of the Hamiltonian, so is $z\psi_n(\xi)$, with z any complex number. Thus we are allowed to assume the initial conditions

$$\begin{aligned} \psi_n(0) &= 1, \qquad & \chi_n(0) &= 0, \qquad & &\text{for } n \text{ even}, \\ \psi_n(0) &= 0, \qquad & \chi_n(0) &= 1, \qquad & &\text{for } n \text{ odd}. \end{aligned} \tag{5.31}$$

Obviously, numerical integration with these initial conditions does not lead to a normalized wavefunction. However, once the wavefunction is found, we can easily evaluate the normalization factor. But we must still determine the correct values of the eigenvalues W_n.

5.5.2 The Shooting Method

The *shooting method* is a method for solving boundary value problems by reducing them to the solution of initial value problems. Roughly speaking, we *shoot out* trajectories in different directions by changing the initial values, until we find the trajectory reaching the desired boundary value. For instance, we know that in our case the boundary conditions for a bound state are

$$\lim_{\xi \to \pm\infty} \psi_n(\xi) = 0. \tag{5.32}$$

These conditions are fulfilled if we insert the conditions (5.31) and the correct value for W_n into the system (5.30), and perform the numerical integration of the ODE system from $\xi = 0$ to $\xi = +\infty$ (actually, up to a sufficiently large value of ξ). We don't need the integration from $\xi = 0$ to $\xi = -\infty$ because of the symmetry of the problem. But, if the value of W_n inserted into (5.30) is not correct, we get a wavefunction $\psi(\xi)$ diverging for $|\xi| \to \infty$. For instance, if we assume $\xi_0 = 10$ in (5.30), the ground state eigenvalue is

$$W_0 = 0.02037903954499\ldots \tag{5.33}$$

(we shall see in the following how this value can be obtained). Figure 5.5 shows the behavior of $\psi(\xi)$ up to $\xi = 30$. If we insert a slightly different value, say $W_0 \pm \delta W$, with $\delta W = 5 \times 10^{-10}$, we see that $\psi(\xi)$ diverges with increasing ξ. In this case we have $\lim_{\xi\to\infty} \psi(\xi) = -\infty$ for $W_0 - \delta W$, and $\lim_{\xi\to\infty} \psi(\xi) = +\infty$ for $W_0 + \delta W$. Obviously a computer does not deal with real numbers, but all our values, including (5.33), are affected by truncation errors. This means that we observe a divergence of ψ also for our "correct" value of W_0, provided that we go up to sufficiently high ξ values. It is a matter of the precision we require, compatibly with the number of bits

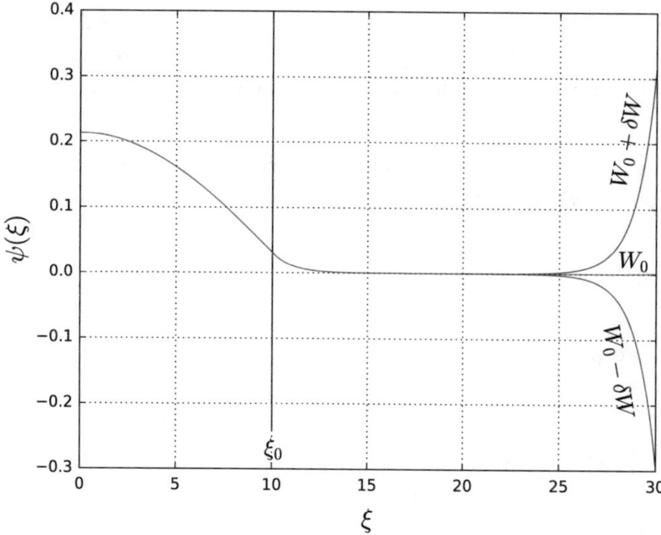

Fig. 5.5 Numerical integrations of the ODE system (5.30), assuming the correct normalized energy eigenvalue W_0 (black line), and the slightly different values $W_0 - \delta W$ (red line) and $W_0 + \delta W$ (blue line), with $\delta W / W_0 \simeq 2.5 \times 10^{-8}$. Incorrect values of W_0 lead to divergence at high ξ

used by the computer for storing numbers. In our case the shooting method works as follows:

1. Not knowing the exact value of W_0, we start by inserting an educated guess $W_0^{(0)} < W_0$ into (5.30). Then we integrate our ODE system up to a sufficiently high value $\xi_{\max}$ of ξ, where we expect $\psi(\xi_{\max})$ to be practically zero. In the present case we know that the energy of our particle must be greater than zero, so, we start from $W_0^{(0)} = 0$. The numerical integration will provide a discretized wavefunction $\psi_0^{(0)}(\xi)$, with $\psi_0^{(0)}(\xi_{\max})$ different from zero and usually very large in absolute value.
2. Now we insert the new value $W_0^{(1)} = W_0^{(0)} + \Delta W$ into (5.30), where the increase ΔW must be smaller than the distance between two consecutive eigenvalues. Since, again, we don't know the eigenvalues and their spacing yet, determining a correct step ΔW requires an educated guess and some trial and error. We integrate the system, and determine $\psi_0^{(1)}(\xi)$. If $\psi_0^{(1)}(\xi_{\max})$ has the same sign as $\psi_0^{(0)}(\xi_{\max})$ we iterate the procedure, with $W_0^{(i+1)} = W_0^{(i)} + \Delta W$, until we find $\psi_0^{(i+1)}(\xi_{\max})\, \psi_0^{(i)}(\xi_{\max}) < 0$.
3. When $\psi_0^{(i+1)}(\xi_{\max})\, \psi_0^{(i)}(\xi_{\max}) < 0$, i.e., when $\psi_0^{(i+1)}(\xi_{\max})$ and $\psi_0^{(i)}(\xi_{\max})$ have opposite sign, the interval $(W_0^{(i)}, W_0^{(i+11)})$ contains the correct ground-state eigenvalue W_0, which can be determined with the required accuracy, for instance, by the bisection method. This will lead to W_0 and $\psi_0(\xi)$.
4. Once W_0 has been determined, we determine the first-excited eigenvalue W_1 and eigenfunction $\psi_1(\xi)$ starting from the initial value $W_1^{(0)} = W_0 + \Delta W$, and

proceeding as in points 1–3. Successive bound eigenstates are determined analogously, always starting with $W_{i+1}^{(0)} = W_i + \Delta W$. Note that the finite square well always has at least one bound state (the ground state), and its number of bound states is finite.

Here follows a possible Python code for determining the bound states of a square well with half-width $\xi_0 = 10$.

Listing 5.2 FiniteWell.py

```
1  #!/usr/bin/env python3
2  #
3  import numpy as np
4  from scipy.integrate import odeint,simps
5  import matplotlib.pyplot as plt
6  #
```

Lines 1–5 are the usual headers.

```
7  xi0=10.0
8  nPointsWell=100
9  nPointsPlot=3*nPointsWell
10 nPoints=5*nPointsWell
11 scale=0.2
12 xiMaxPlot=(xi0*nPointsPlot)/nPointsWell
13 xiMax=(xi0*nPoints)/nPointsWell
14 EigvStep=0.05
15 DeltaXi=xi0/nPointsWell
16 tolerance=1.0e-12;
17 #
18 xi=np.linspace(0,xiMax,nPoints)
19 #
```

Line 7 defines the square-well half-width `xi0` (ξ_0). Lines 8–10 are the numbers of sampling points for the wavefunctions from 0 to the right border of the potential well (`nPointsWell`), from 0 to the right end of the plot (`nPointsPlot`), and the total number of points used for the calculations (`nPoints`). Variable `xiMaxPlot` is the highest plotted ξ value, while `xiMax` is the highest ξ value used for calculations. Variable `EigvStep` is the increase step ΔW, `DeltaXi` is the spacing between consecutive points, and `tolerance` is the required accuracy on the eigenvalues W_n. Line 18 builds the ξ array consisting of `nPoints` evenly spaced values from 0 to ξ_{max} (the points at which the functions must be evaluated).

```
20 def dfdxi(y,xi,params):
21   psi,dpsidxi=y     #   unpack y
22   E,xi0=params      #   unpack parameters
23   if xi<xi0:
24     derivs=[dpsidxi,-E*psi]
25   else:
26     derivs=[dpsidxi,(1-E)*psi]
27   return derivs
28 #
```

Lines 20–28 define the function `dfdxi(y,xi,params)`, which returns the derivatives of ψ and of $\chi = \mathrm{d}\psi/\mathrm{d}\xi$. Array `y` contains the values of ψ and $\mathrm{d}\psi/\mathrm{d}\xi$ at the start point of each integration step, which are unpacked at Line 21. The array `xi` contains the points at which the functions ψ and $\mathrm{d}\psi/\mathrm{d}\xi$ must be evaluated. The array `params` contains the trial eigenvalue $W_n^{(i)}$, unpacked as `E`, and the half-width of the square well ξ_0, unpacked as `xi0`. Lines 23–26 evaluate the derivatives, which are stored into the array `derivs`. The derivative of ψ always equals $\chi = \mathrm{d}\psi/\mathrm{d}\xi$, while we have (Lines 23–26)

$$\frac{\mathrm{d}\chi}{\mathrm{d}\xi} = \frac{\mathrm{d}^2\psi}{\mathrm{d}\xi^2} = \begin{cases} -W_n^{(i)}\psi & \text{if } \xi < \xi_0, \\ \left(1 - W_n^{(i)}\right)\psi & \text{if } \xi > \xi_0. \end{cases} \tag{5.34}$$

Line 27 returns the derivatives to the calling function.

```
29 def SymmWell(params,xi,iEv,EigvStart,EigvStep,tolerance,dfdxi,psi):
30   # ........................................................ initialize
31   eigv1=EigvStart
32   params[0]=eigv1
33   if iEv%2==0:
34     y=[1.0,0.0]
35   else:
36     y=[0.0,1.0]
37   psoln=odeint(dfdxi,y,xi,args=(params,))
38   PsiEnd1=psoln[-1,0]
39   # ............................................... search for interval
40   while True:
41     eigv2=eigv1+EigvStep
42     if eigv2>1.0:
43       return -1
44     params[0]=eigv2
45     psoln=odeint(dfdxi,y,xi,args=(params,))
46     PsiEnd2=psoln[-1,0]
47     if (PsiEnd1*PsiEnd2)<0.0:
48       break
49     PsiEnd1=PsiEnd2
50     eigv1=eigv2
51   # ..................................... logarithmic search for eigenvalue
52   while True:
53     eigvmid=(eigv1+eigv2)/2.0
54     if abs(eigv1-eigv2)<tolerance:
55       break
56     params[0]=eigvmid
57     psoln=odeint(dfdxi,y,xi,args=(params,))
58     PsiEndMid=psoln[-1,0]
59     if (PsiEndMid*PsiEnd1)>0 :
60       PsiEnd1=PsiEndMid
61       eigv1=eigvmid
62     else:
63       PsiEnd2=PsiEndMid
64       eigv2=eigvmid
65   # ................................................. list wave function
```

```
66      del psi[:]
67      for i in range(len(xi)):
68        psi.append(psoln[i,0])
69      # ...............................................................
70      return eigvmid
71    #
```

Lines 29–70 define the function `SymmWell(params, xi, iEv, EigvStart, EigvStep, tolerance, dfdxi, psi)`, which returns the eigenvalue W_n, stored in `eigvmid`, and the corresponding discretized eigenfunction ψ_n, stored in the list `psi`. The arguments `params`, `xi`, `EigvStep`, `tolerance` and `dfdxi` (the function starting at Line 20) have been discussed above. Variable `iEv` is the index n labeling both W_n and ψ_n, needed because $\psi_n(\xi)$ is even if `iEv` is even, odd if `iEv` is odd. Variable `Eigvstart` is the start trial value $W_n^{(0)}$, while `EigvStep` is ΔW, the step used for the search, `psi` is the address of an array where the discretization of ψ_n will be stored.

The lower limit for the eigenvalue search is `eigv1` (corresponding to $W_n^{(0)}$), which is initially set equal to `EigvStart` at Line 31. Line 32 stores `eigv1` into `params(Ø)`, for passing it to function `dfdxi()`. Lines 33–36 set the values of $\psi(0)$ and $\chi(0)$ according to (5.31). Line 37 calls the library function `scipy.integrate.odeint()`, discussed in Sect. 5.3, which integrates a system of ordinary differential equations. Line 39 copies the value of $\psi_n^{(0)}(\xi_{\max})$ into `PsiEnd1`. The loop 40–50 searches for an interval containing the eigenvalue W_n. Line 41 assigns the value $W_n^{(i+1)} = W_n^{(i)} + \Delta W$ to `eigv2`. If $W_n^{(i+1)} > 1$ we are above the well depth ($E > V_0$), and Lines 42–43 break the search, forcing the function to return an impossible negative eigenvalue. Lines 44–46 call `odeint()` again, and evaluate $\psi_n^{(i+1)}(\xi_{\max})$, stored in `PsiEnd2`.

The loop stops if $\psi_n^{(i+1)}(\xi_{\max})\,\psi_n^{(i)}(\xi_{\max}) < 0$, otherwise `PsiEnd2` and `eigv2` are copied into `PsiEnd1` and `eigv1`, and the procedure is iterated. The loop 52–64 evaluates W_n with the required accuracy. Line 53 assigns the arithmetic mean of the upper and lower bounds of the interval to `eigvmid`. If the interval width is smaller than `tolerance` the loop is terminated at Line 55. Otherwise `eigvmid` is stored in `params[Ø]`, `odeint()` is called at Line 57, and the corresponding value of $\psi_n(\xi_{\max})$ is stored into `PsiEndMid`. In Lines 59–64 if `PsiEndMid` and `PsiEnd1` have the same sign, `PsiEndMid` replaces `PsiEnd1`, otherwise `PsiEndMid` repalces `PsiEnd1`, in any case the interval width is halved. At Lines 66–68 list `psi` is cleared, then it is filled with the discretized eigenfunction. At Line 70 `SymmWell()` returns the eigenvalue to the calling function. Here end the function definitions, and the main program starts at Line 73.

```
72  # ...............................................................
73  x=np.linspace(-xiMaxPlot,xiMaxPlot,(2*nPointsPlot)+1)
74  # ...................................................... draw grid
75  plt.grid(True)
76  # ...............................................................
77  eigv=[]
78  EigvStart=0.0;
79  i=0
```

Line 73 stores the ξ interval to be plotted into list x. Line 75 asks for a grid in the plot, Line 77 creates an empty list for storing the eigenvalues, Line 78 assigns the lower end of the search interval of the first eigenvalue, `EigvStart`. Line 79 assigns the index of the ground state, $i = 0$, the eigenfunction will be symmetric.

```
80  while True:
81    params=[EigvStart,xi0]
82    psi=[]
83    eigv.append(SymmWell(params,xi,i,EigvStart,EigvStep,tolerance,dfdxi,psi))
84    if eigv[i]>0:
85      print(i,eigv[i])
86    else:
87      break
88    # ...................................... truncate diverging tail of psi
89    while len(psi)>5:
90      if abs(psi[-2])>abs(psi[-1]):
91        break
92      psi.pop()
93    # ......................................................... normalize psi
94    NormFact=np.sqrt(2.0*simps(np.square(psi),dx=DeltaXi,even='first'))
95    # ............................................... truncate to plot length
96    del psi[(nPointsPlot+1):]
97    if len(psi)<(nPointsPlot+1):
98      while len(psi)<(nPointsPlot+1):
99        psi.append(0.0)
100   normpsi=[i*(scale/NormFact) for i in psi]
101   psineg=list(reversed(normpsi))
102   if i%2==1: # .......................... odd functions are antisymmetric
103     for k in range(len(psineg)):
104       psineg[k]=-psineg[k]
105   # ...................................................... form whole psi
106   psineg.pop()
107   psi=psineg+normpsi
108   #-------------------------------------------------
109   EnerShift=eigv[i]
110   psi=[x+EnerShift for x in psi]
111   plt.plot([-xiMaxPlot,xiMaxPlot],[EnerShift,EnerShift],'black')
112   plt.plot(x,psi)
113   # ...................................................... next eigenvalue
114   EigvStart=eigv[i]+EigvStep
115   i+=1
```

The loop 80–115 searches for the eigenvalues and eigenfunctions of our problem. Line 81 assigns the values of `params` needed by `dfdxi()`, Line 83 calls `SymmWell()` and appends the new energy eigenvalue to `eigv`. The loop is interrupted if `eigv[i]<Ø`, meaning that we have reached the top of the well. If a new eigenvalue has been found the loop 89–93 truncates the list `psi1` where the calculated $\psi(\xi)$ starts to increase because of the round-off error on the eigenvalue. Line 94 evaluates the normalization constant

$$k = \sqrt{2 \int_0^\infty |\psi_n(\xi)|^2 \mathrm{d}\xi}, \tag{5.35}$$

such that $\Psi_n(\xi) = \psi_n(\xi)/k$ is a normalized wavefunction. Actually the upper integration limit is obviously not infinity, but the last point of the list `psi`. Line 96 deletes the list points above the plot limit. If `psi` was already too short, it is zero-padded at Lines 97–99. Line 100 evaluates the normalized wavefunction Ψ_n, times a scale factor needed for the plot. Lines 101–107 build the complete wavefunction, inserting the values corresponding to $\xi < 0$. The function is even if `i` is even, odd if `i` is odd. Line 106 pops the last value of `psineg`, otherwise the value $\psi_n(0)$ would appear twice in the complete list built at Line 107. Lines 109–113 plot the wavefunction, shifted upwards by the energy eigenvalue. Line 111 plots a straight horizontal Line corresponding to the eigenvalue.

Line 114 assigns the lower limit for the search of the next eigenvalue, and Line 115 increases the eigenstate index.

```
116 # ................................................... plot square well
117 plt.plot([-xiMaxPlot,-xi0],[1.0,1.0],color='black')
118 plt.plot([-xi0,-xi0],[1.0,0.0],color='black')
119 plt.plot([xi0,xi0],[0.0,1.0],color='black')
120 plt.plot([xi0,xiMaxPlot],[1.0,1.0],color='black')
121 plt.ylim(0.0,1.1)
122 plt.ylabel(r'Energy/$V_0$',fontsize=18)
123 plt.xlabel(r'$x\,\frac{\sqrt{2mV_0}}{\hbar}$',fontsize=24)
124 plt.tight_layout()
```

Lines 117–124 plot the shape of the square potential well and write the horizontal and vertical axes labels.

```
125 #———————————————————————————————————————————————————————
126 plt.savefig('SquareWell00.eps', format='eps', dpi=1000)
127 plt.show() # show the plot
```

Line 126 saves the plot in encapsulate PostScript format (.eps), and Line 127 shows the plot on the computer monitor. The result is shown in Fig. 5.6. The horizontal black lines correspond to the energy eigenvalues divided by V_0, namely

$$\begin{array}{ll} W_0 & 0.020379039559702515 \\ W_1 & 0.08135854295251191 \\ W_2 & 0.18242255589320847 \\ W_3 & 0.32253401889174704 \\ W_4 & 0.4996922434624138 \\ W_5 & 0.7095036776365305 \\ W_6 & 0.9368079801646677 \end{array} \tag{5.36}$$

these values are printed on the computer monitor by Line 85. As we know from quantum mechanics, a finite square well always has at least one bound state (the ground state). The number of bound excited states depends on the depth and on the width of the well. In our case, $\xi_0 = x_0\sqrt{2mV_0}/\hbar = 10$, we have seven bound

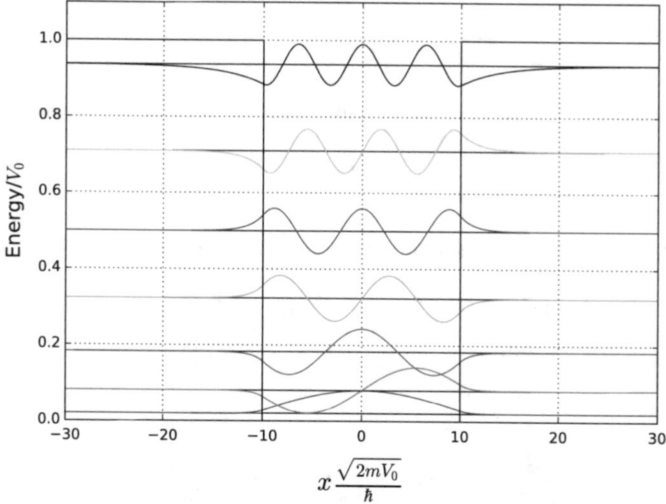

Fig. 5.6 The seven bound eigenstates of a finite potential well of half-width $x_0 = 10\,\hbar/\sqrt{2mV_0}$, i.e., $\xi_0 = 10$. The ground state is always bound, whatever the width of the well. The number of excited bound eigenstates increases with increasing ξ_0

states, six excited states plus the ground state. Eigenstates of higher energy are not bound, but free. The number of bound states increases with increasing ξ_0, i.e., with increasing x_0 at constant V_0, with increasing V_0 at constant x_0, or with both x_0 and V_0 increasing.

5.5.3 The Harmonic Oscillator

Contrary to the square well, the quantum harmonic oscillator has an exact analytical solution. All eigenstates are bound, with energy eigenvalues $E_n = (n + 1/2)\,\hbar\omega$, and Hermite polynomials as eigenfunctions. Here, however, we shall solve the problem by numerical integration combined with the shooting method, and compare our results to the analytical solutions.

The potential energy for the harmonic oscillator is $V(x) = m\omega^2 x^2/2$, where m is the mass of the "oscillating" particle, and $\omega = \sqrt{k/m}$, k being the Hooke constant. The corresponding Schrödinger equation is thus

$$\left(\frac{m\omega^2}{2}x^2 - \frac{\hbar^2}{2m}\frac{\mathrm{d}^2}{\mathrm{d}x^2}\right)\psi_n(x) = E_n\psi_n(x). \tag{5.37}$$

Equation (5.37) can be rewritten in the following form, suitable for ODE numeric integration

$$\frac{\hbar^2}{2m}\frac{\mathrm{d}^2\psi_n(x)}{\mathrm{d}x^2} = \left(\frac{m\omega^2}{2}x^2 - E_n\right)\psi_n(x). \tag{5.38}$$

In order to write (5.38) in terms of dimensionless quantities, we introduce the new variable

$$\xi = \frac{x}{\alpha}, \quad \text{so that} \quad x = \alpha\xi, \tag{5.39}$$

where α is a quantity with the dimensions of a length, to be chosen later. We obtain

$$\frac{\hbar^2}{2m\alpha^2}\frac{\mathrm{d}^2\psi_n(\xi)}{\mathrm{d}\xi^2} = \frac{m\alpha^2\omega^2}{2}\left(\xi^2 - \frac{2}{m\alpha^2\omega^2}E_n\right)\psi_n(\xi). \tag{5.40}$$

We divide both sides by $m\alpha^2\omega^2/2$,

$$\frac{\hbar^2}{m^2\omega^2\alpha^4}\frac{\mathrm{d}^2\psi_n(\xi)}{\mathrm{d}\xi^2} = \left(\xi^2 - \frac{2}{m\alpha^2\omega^2}E_n\right)\psi_n(\xi). \tag{5.41}$$

and we choose α as

$$\alpha = \sqrt{\frac{\hbar}{\omega m}}, \tag{5.42}$$

which has the dimensions of a length, as required. The equation reduces to

$$\frac{\mathrm{d}^2\psi_n(\xi)}{\mathrm{d}\xi^2} = \left(\xi^2 - \frac{2}{\hbar\omega}E_n\right)\psi_n(x). \tag{5.43}$$

If we further define the quantity

$$W_n = \frac{E_n}{\hbar\omega}, \tag{5.44}$$

which is the nth energy eigenvalue measured in units of $\hbar\omega$, our final equation involves only dimensionless quantities, and is written

$$\frac{\mathrm{d}^2\psi_n(\xi)}{\mathrm{d}\xi^2} = \left(\xi^2 - 2W_n\right)\psi_n(x). \tag{5.45}$$

Equation (5.45) is equivalent to the following system of two first-order ordinary differential equations

$$\begin{aligned} \frac{\mathrm{d}\psi_n}{\mathrm{d}\xi_n} &= \chi_n, \\ \frac{\mathrm{d}\chi_n}{\mathrm{d}\xi} &= \left(\xi^2 - 2W_n\right)\psi_n(x). \end{aligned} \tag{5.46}$$

As in the case of the finite square well, if the correct eigenvalues W_n, and the corresponding appropriate initial values $\psi_n(0)$ and $\chi_n(0)$ are introduced into (5.46), numerical integrations lead to the correctly discretized $\psi_n(\xi)$ and $\chi_n(\xi) = \mathrm{d}\psi_n(\xi)/\mathrm{d}\xi$ functions. Again, the potential energy is an even function of x (and of ξ), implying that the eigenfunctions $\psi_n(\xi)$ are either even, thus with $\chi_n(0) = \mathrm{d}\psi_n(0)/\mathrm{d}\xi = 0$, or odd, with $\psi_n(0) = 0$. The eigenfunction corresponding to the ground state, $\psi_0(\xi)$, is even. Thus we assume again the initial conditions (5.31), and, again, we determine the eigenvalues and eigenfunctions by the shooting method. The harmonic oscillator has infinite eigenvalues, in the following we shall confine ourselves to the lowest seven states, $n = 0, 1, 2, \ldots, 6$.

Listing 5.3 QuantOscill.py

```
1  #!/usr/bin/env python3
2  #
3  import numpy as np
4  from scipy.integrate import odeint,simps
5  import matplotlib.pyplot as plt
6  plt.rc('text', usetex=True)
7  #
8  nPoints=500
9  nPointsPlot=200
10 xiMax=10.0
11 xiMaxPlot=(xiMax*nPointsPlot)/nPoints
12 DeltaXi=xiMax/float(nPoints)
13 scale=4.0
14 nEigen=7
15 EigvStep=0.005
16 tolerance=1.0e-12;
17 #
```

Variable `nPoints` is the number of points used for calculations, `nPointsPlot` is the number of points displayed in the plot, `xiMax` is the maximum ξ value, $\xi_{\max}$, used for calculations, it is assumed that $\psi_n(\xi_{\max})$ is practically zero for the evaluated functions (up to $n = 6$), `xiMaxPlot` is the maximum xi value displayed in the plot. `DeltaXi` is the ξ spacing between two consecutive points, needed for evaluating the integral of $|\psi|^2$ used for normalizing ψ. Quantity `nEigen` is the number of evaluated eigenstates, 6 plus the ground state in our case: since the bound eigenstates of the present problem are infinite, we must set a limit somewhere!

```
18 xi=np.linspace(0,xiMax,nPoints)
19 #
20 def dfdxi(y,xi,params):
21     psi,dpsidt=y     #  unpack y
22     E,=params        #  unpack parameters
23     derivs=[dpsidt,(xi*xi-2.0*E)*psi]
24     return derivs
25 #
```

Analogous to the square well. Here the tentative eigenvalue `E` is the only parameter needed by the function `dfdxi()`, Line 23 evaluates the derivative $\mathrm{d}\chi/\mathrm{d}\xi$ according to (5.46).

```
def SymmWell(params,xi,iEv,EigvStart,EigvStep,tolerance,dfdxi,psi):
  # ........................................................ initialize
  eigv1=EigvStart
  params[0]=eigv1
  if iEv%2==0:
    y=[1.0,0.0]
  else:
    y=[0.0,1.0]
  psoln=odeint(dfdxi,y,xi,args=(params,))
  PsiEnd1=psoln[-1,0]
  # ................................................ search for interval
  while True:
    eigv2=eigv1+EigvStep
    params[0]=eigv2
    psoln=odeint(dfdxi,y,xi,args=(params,))
    PsiEnd2=psoln[-1,0]
    if (PsiEnd1*PsiEnd2)<0.0:
      break
    PsiEnd1=PsiEnd2
    eigv1=eigv2
  # ................................. logarithmic search for eigenvalue
  while True:
    eigvmid=(eigv1+eigv2)/2.0
    params[0]=eigvmid
    if abs(eigv1-eigv2)<tolerance:
      break
    psoln=odeint(dfdxi,y,xi,args=(params,))
    PsiEndMid=psoln[-1,0]
    if (PsiEndMid*PsiEnd1)>0 :
      PsiEnd1=PsiEndMid
      eigv1=eigvmid
    else:
      PsiEnd2=PsiEndMid
      eigv2=eigvmid
  # ................................................ list wave function
  del psi[:]
  for i in range(len(xi)):
    psi.append(psoln[i,0])
  # ...................................................................
  return eigvmid
#
```

Analogous to the corresponding code for the finite square well.

```
# ........................................ evaluate and draw potential
x = np.linspace(-xiMaxPlot,xiMaxPlot,(2*nPointsPlot)+1)
y = 0.5*x**2 # potential.
plt.plot(x,y) # x^2
# ......................................................... draw grid
plt.grid(True)
# ..................................................................
```

These lines draw the harmonic potential on the plot. The potential is

$$V(x) = \frac{1}{2}\,\omega^2 m\, x^2 \quad \Rightarrow \quad V(\xi) = \frac{1}{2}\,\hbar\omega\,\xi^2 \quad \Rightarrow \quad V(\xi) = \frac{1}{2}\,\xi^2, \tag{5.47}$$

the last formula being in units of $\hbar\omega$. The rest is analogous to the code for the finite square well.

```
74 eigv=[]
75 EigvStart=0.0;
76 i=0
```

Analogous to the finite square well.

```
77 while i<nEigen:
78   params=[EigvStart]
79   psi=[]
80   eigv.append(SymmWell(params,xi,i,EigvStart,EigvStep,tolerance,dfdxi,psi))
81   print(i,eigv[i])
82   # ........................................ truncate diverging tail of psi
83   while len(psi)>5:
84     if abs(psi[-2])>abs(psi[-1]):
85       break
86     psi.pop()
87   # ........................................................... normalize psi
88   NormFact=np.sqrt(2.0*simps(np.square(psi),even='first'))
89   # .............................................. truncate to plot length
90   del psi[(nPointsPlot+1):]
91   if len(psi)<(nPointsPlot+1):
92     while len(psi)<(nPointsPlot+1):
93       psi.append(0.0)
94   normpsi=[i*(scale/NormFact) for i in psi]
95   psineg=list(reversed(normpsi))
96   if i%2==1: # ........................... odd functions are antisymmetric
97     for k in range(len(psineg)):
98       psineg[k]=-psineg[k]
99   # ....................................................... form whole psi
100  psineg.pop()
101  psi=psineg+normpsi
102  #———————————————————————————————
103  EnerShift=eigv[i]
104  psi=[x+EnerShift for x in psi]
105  plt.plot([-xiMaxPlot,xiMaxPlot],[EnerShift,EnerShift],'black')
106  plt.plot(x,psi)
107  # ...................................................... next eigenvalue
108  EigvStart=eigv[i]+EigvStep
109  i+=1
```

Lines 77–109 constitute the main loop of the program, analogous to the main loop of the square well. The only important difference is that the number of bound eigenstates of the harmonic oscillator is infinite, thus Line 77 sets the number of required eigenstates. The eigenfunctions are plotted in Fig. 5.7.

```
# ..................................................................
plt.ylabel('$W=E/(\hbar\omega)$',fontsize=18)
plt.xlabel('$\\displaystyle\\xi=\\sqrt{\\frac{\\omega m}{\\hbar}}\\,x$',fontsize=24)
plt.tight_layout()
# ..................................................................
plt.savefig('QuantOscill00.eps', format='eps', dpi=1000)
plt.show() # show the plot
```

The numerically evaluated eigenvalues are, in units of $\hbar\omega$

$$\begin{array}{ll} W_0 & 0.4999999798237699 \\ W_1 & 1.4999999905895576 \\ W_2 & 2.500000022660503 \\ W_3 & 3.4999999952432175 \\ W_4 & 4.500000038020879 \\ W_5 & 5.499999974849539 \\ W_6 & 6.500000020207548, \end{array} \tag{5.48}$$

they approximate within some 10^{-8} the analytically calculated values

$$W_n = n + \frac{1}{2} \tag{5.49}$$

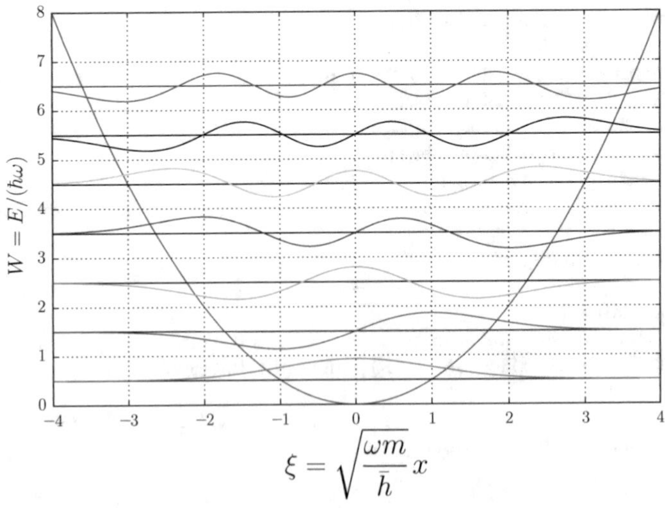

Fig. 5.7 The ground state and the first six excited eigenstates of the quantum harmonic oscillator, as obtained by numerical integration

The version of the shooting method discussed in the last sections of this chapter can be adapted to any symmetric one-dimensional attractive potential. Notable cases are the ammonia inversion potential and the oxetane ring-puckering potential in molecular physics.

5.6 Exercises

Exercise 5.1 Assume that the bob of the pendulum of Sect. 5.4 has mass $m = 5 \times 10^{-2}$ kg and is subject to a drag force $\boldsymbol{f}_{\text{drag}} = -\beta\, \boldsymbol{v}$, where $\beta = 10^{-2}$ kg/s is a constant, and $\boldsymbol{v}$ is the bob velocity. Modify Listing 5.1 in order to take drag into account and plot the pendulum angle $\vartheta(t)$ in the interval $0 \le t \le 12$ s. Disregard the comparison with the pure harmonic motion.

Exercise 5.2 A spherical cluster comprising $N = 10^8$ helium atoms has initial radius $r_0 = 10^{-7}$ m. At time $t = 0$ all the atoms, initially at rest, are ionized by an intense laser pulse. Assume that each He atom is singly ionized, acquiring a positive charge $q = 1.6 \times 10^{-19}$ C. Since the ions repel one another, a "Coulomb explosion" follows. It can be shown that, during the explosion, the atoms do not overtake one another, and spherical symmetry is preserved, see Problem 1.6 of Reference [1]. Regarding the charge distribution as continuous, plot the radius of the cluster as a function of time in the interval $0 < t < 5 \times 10^{-9}$ s. Assume the value $k_\text{e} = 1/(4\pi\varepsilon_0) \approx 9 \times 10^9$ $\text{N}\,\text{m}^2\,\text{C}^{-2}$ for the Coulomb constant, and $m = 4\ \text{u} \approx 6.64 \times 10^{-27}$ kg for the mass of the He ion.

Exercise 5.3 A long, straight, uniformly charged wire has linear charge density $\lambda = -2.0 \times 10^{-9}$ C/m. A proton (charge $q = 1.602 \times 10^{-19}$ C, mass $m = 1.673 \times 10^{-27}$ kg) is orbiting about the wire in a perpendicular plane. At time $t = 0$ the proton is at a distance $r_0 = 10^{-2}$ m from the wire, and its velocity is perpendicular to the distance wire-proton.

(a) Evaluate the value v_c of the velocity corresponding to a circular orbit about the wire. Assume the value $\varepsilon_0 = 8.854 \times 10^{-12}$ F/m for vacuum permittivity.
(b) Only circular orbits are closed in a logarithmic potential. Assume that the proton velocity at $t = 0$ is $v_0 = v_\text{c}/2$. Write a program that plots the proton trajectory from $t = 0$ to $t = 5 \times 10^{-6}$ s. Hint: divide the time interval in subintervals of 10^{-8} s each.

Reference

1. A. Macchi, G. Moruzzi, F. Pegoraro, *Problems in Classical Electromagnetism* (Springer, Berlin, 2017)

Chapter 6
Tkinter Graphics

6.1 Tkinter

A *graphical user interface* (GUI) is a user interface that allows users to interact with a computer by clicking the mouse on graphical icons representing, for instance, buttons or menus, rather than by typing commands at the command line. As with all other computer graphical tools, for long time a problem with GUIs has been the lack of cross-platform compatibility, i.e., the impossibility to use the same GUI on different operating systems, notably Linux, macOS and Windows.

Tcl (suggested pronunciation: "tickle") is a high-level programming language designed for being very simple but powerful. The most popular Tcl extension is the *Tk* toolkit, first announced in 1991, which provides a graphical user interface library for a variety of operating systems, thus achieving a wide cross-platform compatibility. The popular combination of Tcl with the Tk extension is referred to as Tcl/Tk, and enables building a GUI natively in Tcl. Tcl/Tk is included in the standard Python installation in the form of *Tkinter*, standing for "Tk interface". Tkinter, which we are going to consider in this chapter, is Python's de facto standard GUI. Tkinter is included with the standard Linux, Microsoft Windows and macOS installations of Python.

There are several popular GUI library alternatives available, such as wxPython, PyQt (PySide), Pygame, Pyglet, and PyGTK, which, however, we are not going to consider here.

We shall still use the command line or Spyder for preparing our Tkinter based scripts, but, once our scripts are running, we interact with the program execution through the Tkinter GUI. Further, what is perhaps more important, Tkinter provides more freedom than Pyplot both in drawing figures and in animation on the computer monitor. In the present chapter we shall discuss the basics of Tkinter graphics, while Tkinter animation will be considered in Chap. 7.

Electronic supplementary material The online version of this chapter (https://doi.org/10.1007/978-3-030-45027-4_6) contains supplementary material, which is available to authorized users.

G. Moruzzi, *Essential Python for the Physicist*,
https://doi.org/10.1007/978-3-030-45027-4_6

6.2 The Root Window and Tkinter Color Management

We start by describing how Tkinter manages the colors we use for graphics. Tkinter represents colors with strings. There are two ways to specify colors in Tkinter:

1. You can use locally predefined standard color names. The list is platform dependent, and you can inquire on the internet what is available for your particular platform. However, the color strings `'white'`, `'black'`, `'red''`, `'green'`, `'blue'`, `'cyan'`, `'yellow'`, and `'magenta'` are always available. This is enough for most applications.
2. You can define your own colors by using strings specifying the intensities of the three primary colors *red*, *green* and *blue* in hexadecimal digits, according to the RGB additive color model. The intensity of each primary color is represented by one ore two hexadecimal digits, according to the color depth. The relative intensity of each primary color is 0 – f (decimal 0 – 15) if a single hexadecimal digit is used, 00 – ff (decimal 0 – 255) if two digits are used. The full color string comprises a hash symbol (#) followed either by three or six hexadecimal digits. For example, in the 6-hexadecimal-digits representation, "#ffffff" corresponds to white (maximum intensity, 255, or ff in hexadecimal, for all three primary colors), "#000000" to black (minimum intensity, 0, for all primary colors), "#ff0000" to pure bright red, "#010000" to the darkest possible red, "#00ff00" to pure bright green, "#0000ff" to pure bright blue, and "#00ffff" to bright cyan (green plus blue, both at maximum intensity).

Listing 6.1 is a simple script that you can use to test the correspondence between strings and colors on your computer monitor. You can run it from the command line by typing, for instance
`$>checkcolor.py e0cc0f`
the string `e0cc0f` asks for the relative color intensities (in decimal), red=224, green=204, blue=15. If you use Spyder3 instead of the command line, once you have Script 6.1 in the editor subwindow, you first click on the Run option on the top menu bar. As in Sect. 2.1.3, this opens a menu window where you click on Command line options:, then you type `e0cc0f` in the window at the right of Command line options:. Finally you click on the Run option at the bottom of the menu window.

Listing 6.1 checkcolor.py

```
#!/usr/bin/env python3
from tkinter import Tk,Canvas
from sys import argv
#
script,col=argv
colstring=''#''+col
root=Tk()
root.title('Check Color')
canvas=Canvas(root,width=200,height=200,background=colstring)
canvas.grid(row=0,column=0)
#
root.mainloop()
```

Fig. 6.1 Output of Listing 6.1.

Functions `Tk()` and `Canvas()` are imported from Tkinter at Line 2, while the list `argv` is imported from the `sys` library at Line 3. As we already know, list `argv` comprises everything we typed on the the command line, so `checkcolor.py` is copied into the string `script`, and `e0cc0f` is copied into the string `col` at Line 5. Thus, string `colstring` at Line 6 becomes `#e00cc0f` in our case. Line 7 creates the root window (which is always needed when using Tkinter graphics, and is the only window we shall use in this program), and Line 8 writes the title *Check Color* in the frame of the root window, as shown in Fig. 6.1. Line 9 creates a *canvas*, the surface where we can draw and paint, 200 pixel wide and 200 pixel high in the root window: Tkinter measures lengths in *pixels* (abbreviated as px). The background color of the canvas is set to the color coded in `colstring`. In Line 10 we are introduced to the Tkinter `grid()` geometry manager. Just think of the root window as divided into contiguous rectangular surfaces organized in rows and columns. Rows and columns are numbered from zero upwards. Since here the canvas is the only widget present in the root window, it is located at row=0, column=0. Note that writing, for instance, `canvas.grid(15,27)` would not alter the output, since empty rows and columns are simply ignored by Tkinter. However, Line 10 *must* be in the script, otherwise Python does not know where to locate the canvas. Just see what happens if you cancel it out. We shall see the importance of the grid geometry manager in Sect. 7.3, Listing 7.3. Line 12 starts the execution of the program, and the root window frame and the colored canvas are displayed. The only way to close the program is to click the mouse on the `x` in the small red circle at the upper left of the frame. Later we shall learn more "refined" ways to exit programs. You can experiment all possible color encodings. The result is shown in Fig. 6.1. Also note that, if you type

```
$>checkcolor.py
```

on the terminal command line, without the color code, Python will complain that something is missing in the command line.

As a further example of the use of the color codes, Listing 6.2 draws an approximate rainbow

Listing 6.2 Rainbow.py

```
1  #!/usr/bin/env python3
2  from tkinter import Tk,Canvas
3  #
4  root=Tk()
5  root.title('Rainbow')
6  canvas=Canvas(root,width=800,height=150,background="#ffffff")
7  canvas.grid(row=0,column=0)
8  #
```

The code above is analogous to the code of Lines 1–10 of Listing 6.1, the only difference being that the list `sys.argv` is not needed here. Line 6 creates a canvas 800 pixels wide and 150 pixels high belonging to the root window, and the background color is set to white.

```
9  for i in range(0,800):
10   ColString="#"
11   if i<256:
12     r=255;g=i;b=0;
13   elif i<512:
14     r=511-i;g=255;b=i-256;
15   else:
16     r=0;g=767-i;b=255;
17     if i>672:
18       r=(i-672)*2
19     if g<0:
20       g=0
21 #
22   ColString=ColString+format(r,'02x')+format(g,'02x')+format(b,'02x')
23   line=[i,0,i,150]
24   canvas.create_line(line,fill=ColString)
25 #
26 root.mainloop()
```

The loop 9–24 paints 800 vertical lines, numbered by the index `i`, each line one-pixel wide, in the canvas. Each line will have a different color, according to its index `i`. The integer variables `r`, `g` and `b` represent the intensities of the primary colors red, green and blue. Line 10 creates a one-character string containing the initial hash symbol `#`. For $0 \le i < 256$ quantity `r` is assigned the maximum value 255 (maximum red intensity), `b` is always 0, while `g` gradually increases from 0 to 255. For $256 \le i < 512$ the value of `r` decreases from 255 to 0, `g` has always its maximum value 255 (maximum green intensity), and `b` increases from 0 to 255. For $i \ge 512$ the value of `r` is always 0, `g` decreases from 255 to 0, and `b` has always its maximum value 255. For $672 < i < 800$ we have `g` $= 0$, `b` $= 255$, while `r` increases from 2 to 127, in order to reproduce violet. The color vision in the human eye is due to the presence of three types of photoreceptor cells in the retina, called *cones*. The three types of cones have different response curves to the light frequency. The first type responds the most to light of long wavelengths, peaking at about 560 nm, corresponding to red. The second type responds the most to light of medium-wavelength, peaking at 530 nm, corresponding to green. The third type responds the most to short-wavelength light, peaking at 420 nm, corresponding to blue. However, the red-sensible cones are

Fig. 6.2 The "rainbow" output of Listing 6.2

slightly excited also by the higher end of the visible radiation, where photons are more energetic. This is why some red is needed to simulate our visual perception of violet. The method `.create_line()` used at Line 24 is discussed in Sect. 6.3. The output of script 6.2 is shown in Fig. 6.2.

6.3 Drawing Geometric Shapes on the Canvas

Tkinter has its `.create` methods for drawing geometrical shapes and writing text on the canvas. Geometric shapes and text are positioned on the canvas using the coordinate system shown in Fig. 6.3. As usual when dealing with displays on a computer monitor, the coordinate origin is located at the upper left corner of the window, and the y axis is directed downwards. Both the x and the y coordinates are measured in pixels.

We have already met the method `.create_line()` at Line 24 of Listing 6.2, where it was used to draw vertical lines of different colors. Listing 6.3 illustrates the use of the further methods `.create_rectangle()`, `.create_polygon()`, `.create_oval()`, `.create_arc()` and `.create_text()`, see Fig. 6.4. All these methods have both mandatory and optional arguments. See Appendix D fot a more complete discussion.

Listing 6.3 GeomShapes.py

```
#!/usr/bin/env python3
from tkinter import Tk,Canvas,ARC,CHORD,PIESLICE
from numpy import cos,sin,pi
#
cw=600
ch=500
```

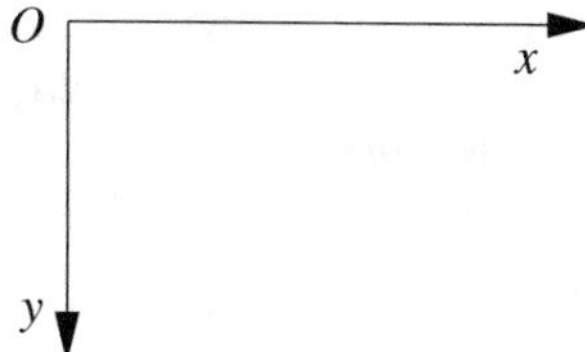

Fig. 6.3 The x and y axis on the Tkinter canvas

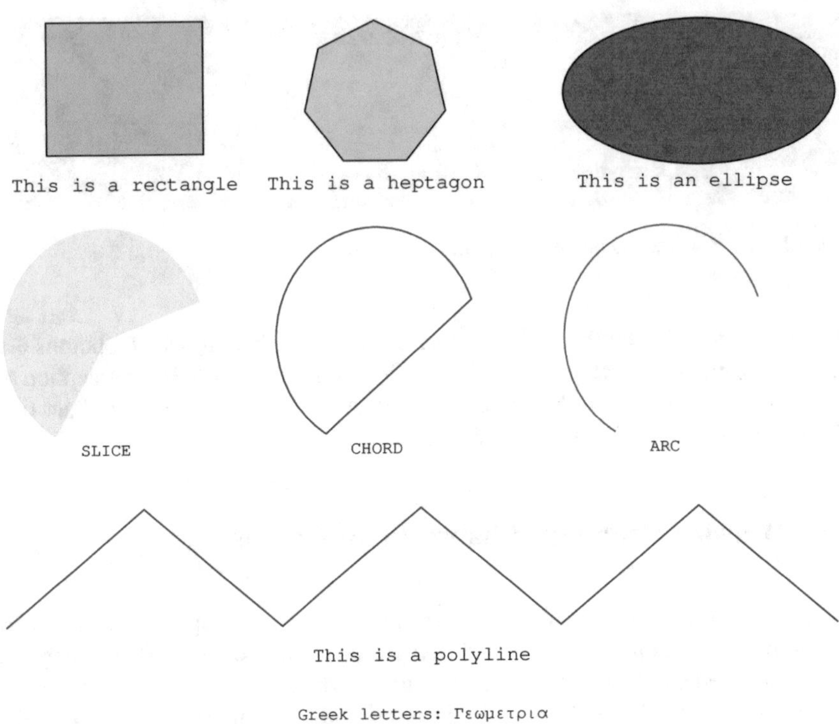

Fig. 6.4 Drawing geometrical shapes and writing text on the canvas

```
7  root=Tk()
8  root.title('Geometric Shapes')
9  canvas=Canvas(root,width=cw,height=ch,background="#ffffff")
10 canvas.grid(row=0,column=0)
```

Variables `cw` and `ch` are the canvas width and height, respectively. Lines 7–10 create the root window, and create and locate the canvas.

```
11 # ................................................... rectangle
12 canvas.create_rectangle(40,10,150,100,fill='#00ff00')
13 canvas.create_text(95,120,text='This is a rectangle',\
14    font=('Helvetica','14'))
```

The method `.create_rectangle()` at Line 12 creates a rectangle with opposite vertices ($x_1 = 40$ px, $y_1 = 10$ px) and ($x_2 = 150$ px, $y_2 = 100$ px), using light green, '#00ff00', as fill color. The mandatory arguments of this method are the four x_1, y_1, x_2 and y_2 coordinates. Here we are using the optional argument `fill='#00ff00'`, specifying the fill color. If you omit this argument, the default color is *transparent*, equivalent to `fill=''`. Another optional argument is `outline`, determining the color of the border. The default is `outline='black'`.

At Lines 13–14 the method `.create_text()` writes a text on the canvas at a position determined by its two mandatory arguments, x and y. The text to be written is specified by the optional argument `text='This is a rectangle'`. By default, the text is printed centered around the (x, y) position. Different positioning with respect to (x, y) is possible via the optional argument `anchor`, the default is `anchor=CENTER`. The optional argument `font` is a tuple comprising two strings, specifying the font, here `'Helvetica'`, and the font size, here `'14'`.

```
15 # ....................................................... heptagon
16 Ox=270
17 Oy=60
18 r=50.0
19 Np=7
20 poly=[]
21 i=0
22 alpha=2.0*pi/Np
23 while i<Np:
24     poly.append(Ox+r*sin(i*alpha))
25     poly.append(Oy-r*cos(i*alpha))
26     i+=1
27 canvas.create_polygon(poly,fill='#00ffff',outline='#000000')
28 canvas.create_text(270,120,text='This is a heptagon',\
29     font=('Helvetica','14'))
```

Here we create a regular polygon, i.e., a closed polyline of n line segments of equal length and n vertices, with $n = 7$ (a heptagon) in the present case. Our heptagon is centered at $O_x = 270$ px, $O_y = 60$ px, and inscribed in a circle of radius $r = 50$ px. Line 20 creates the empty list `poly`, that will contain the vertices of the heptagon. Variable `alpha` defined at line 22 is the central angle, and the loop 23–26 fills the list of the vertex coordinates. Method `.create_polygon()` at Line 27 creates a *polygon* whose vertices are specified by the (x, y) coordinates contained in the list `poly`, the mandatory argument. The optional argument `fill='#00ffff'` specifies that the color of the polygon surface is bright cyan, while the other optional argument `outline='#000000` specifies that the border is black.

```
30 # ....................................................... ellipse
31 canvas.create_oval(400,10,590,109,fill='red')
32 canvas.create_text(485,120,text='This is an ellipse',\
33     font=('Helvetica','14'))
```

Method `create_oval()` draws an ellipse inscribed in a rectangle of opposite vertices (400, 10) and (590, 109). The fill color is `'red'`, the default color, `'black'`, is used for the border since the `outline` option is not specified.

```
34 # ....................................................... arc
35 canvas.create_arc(10,150,150,300,start=20,extent=220,fill='#ffff00',\
36     outline='#ffff00',style=PIESLICE)
37 canvas.create_text(80,300,text='SLICE',font=('Helvetica','12'))
38 canvas.create_arc(200,150,340,300,start=20,extent=220,fill='',\
39     outline='#0000ff',style=CHORD)
40 canvas.create_text(270,300,text='CHORD',font=('Helvetica','12'))
41 canvas.create_arc(400,150,540,300,start=20,extent=220,outline='#ff0000'\
```

```
42      ,style=ARC)
43  canvas.create_text(470,300,text='ARC',font=('Helvetica','12'))
```

Method `create_arc()` is called with different options. Option `start` is the start angle, in degrees, of the arc. The angle is measured from the $+x$ direction, counterclockwise. Option `extent` is the angular width of the arc, again in degrees. Option `style=PIESLICE` draws a *slice* of a pie chart, `style=CHORD` draws a chord connecting the endpoints of the arc, and `style=ARC` simply draws the arc. Constants `ARC`, `CHORD` and `PIESLICE` are defined in the TKinter library and imported at Line 2.

```
44  # ....................................................... line
45  line=[]
46  i=0
47  dx=(cw-20)/6
48  dy=40
49  while i<7:
50      line.append(10+i*dx)
51      line.append(380+dy)
52      dy=-dy
53      i+=1
54  canvas.create_line(line,fill='blue')
55  canvas.create_text(300,440,text='This␣is␣a␣polyline',\
56      font=('Helvetica','14'))
```

Method `create_line()` at Line 54 draws a polyline comprising an arbitrary number n of line segments, obviously including the single segment as special case, and $n+1$ vertices. The mandatory argument `line` is a list comprising the coordinates of the polyline vertices in the form $[x_0, y_0, x_1, y_1, \ldots, x_n, y_n]$. The optional argument `fill` specifies the line color. List `line` is created at lines 45–53.

```
57  # ............................................... Greek letters
58  canvas.create_text(300,480,text=\
59      'Greek␣letters:␣\
60  \u0393\u03B5\u03c9\u03BC\u03B5\u03C4\u03C1\u03B9\u03B1',\
61      font=('Helvetica','12'))
62  #
63  root.mainloop()
```

The `create_text()` command at lines 58–61, split into four lines because of page size, shows the use of Greek letters. Unfortunately it is not (yet?) possible to insert LaTeX text into Tkinter, as we did under Matplotlib. The only way out is using the UTF-8 encoded Greek characters, listed in Table E.2 of Appendix E.

6.4 Plotting a Function with Tkinter

6.4.1 Plotting a Hyperbola

Plotting a function with Tkinter is slightly less easy than with Matplotlib, but still straightforward. Plotting relies on the method `canvas.create_line()`, which draws a polyline. If the line segments of the polyline are sufficiently short, the polyline is indistinguishable from a smooth curve: remember that, in any case, you cannot have a resolution better than 1 px on the computer monitor.
As an example, Listing 6.4 draws a hyperbola of equation

$$\frac{x^2}{a^2} - \frac{y^2}{b^2} = 1\,, \tag{6.1}$$

with $a = 80$ and $b = 40$, on the canvas. The pixel is used as unit of length.

Listing 6.4 HyperbolaPlot.py

```
1  #!/usr/bin/env python3
2  from tkinter import Tk,Canvas,E,LAST,N
3  import numpy as np
4  #
5  cw=800
6  ch=400
7  Ox=cw/2
8  Oy=ch/2
9  # ..................................... create root window and canvas
10 root=Tk()
11 root.title('Hyperbola Plot')
12 canvas=Canvas(root,width=cw,height=ch,background="#ffffff")
13 canvas.grid(row=0,column=0)
```

In addition to the functions `Tk()` and `Canvas()`, we import also the Tkinter constants `E`, `LAST` and `N`. The canvas size is 800 × 400 pixels, `Ox` and `Oy` are the coordinates of the origin of our xy reference frame relative to the canvas reference.

```
14 # ............................................................. axes
15 canvas.create_line(0,Oy,cw-1,Oy,fill='black',arrow=LAST,
16                    arrowshape=(20,20,5))
17 canvas.create_line(Ox,ch-1,Ox,0,fill='black',arrow=LAST,
18                    arrowshape=(20,20,5))
19 canvas.create_text(cw-20,Oy+11,text='x',font=('Times','16',
20                    'italic'))
21 canvas.create_text(Ox-15,15,text='y',font=('Times','16',
22                    'italic'))
```

Lines 15–16 and 17–18 draw the x and y axes on the canvas, respectively. The two axes have arrows at their terminal endpoints (`arrow=LAST`), the shape of the arrows is specified by the `arrowshape` optional parameter, see Appendix D. Lines 19–20 and 21–22 write the symbols x and y at the axes ends.

```
23 # ........................................................ x-ticks
24 dx=80
25 i=1
26 x=0
27 while x<cw-dx:
28    x=i*dx
29    canvas.create_line(x,Oy,x,Oy+10)
30    canvas.create_text(x,Oy+10,text=str(x-Ox),anchor=N)
31    i+=1
32 # ........................................................ y-ticks
33 dy=40
34 i=1
35 y=0
36 while y<ch-dy:
37    y=i*dy
38    txt=str(y-Oy)
39    canvas.create_text(Ox-10,ch-y,text=str(y-Oy),anchor=E)
40    i+=1
```

Lines 24–40 draw the ticks along the x and y axes, respectively. The x ticks are spaced by 80 px, the y ticks by 40 px. The arguments of the method `.create_text()` are discussed in Appendix D.

```
41 # .................................................... asymptotes
42 canvas.create_line(0,0,cw-1,ch-1,fill='red')
43 canvas.create_line(0,ch-1,cw-1,0,fill='red')
```

Lines 42–43 draw the hyperbola asymptotes, the two red diagonals in Fig. 6.5.

```
44 # ................................ make lists of hyperbola coordinates
45 a=80.0
46 b=40.0
47 y=-ch/2
48 hyp1=[]
49 hyp2=[]
50 while y<ch/2:
51    x=(a/b)*np.sqrt(y*y+b*b)
52    hyp1.append(Ox+x)
53    hyp1.append(Oy-y)
54    hyp2.append(Ox-x)
55    hyp2.append(Oy-y)
56    y+=2
```

The equation of our hyperbola is (6.1). The parameters a and b are defined at lines 45 and 46, all values are in pixels. Line 47 sets the initial value of y as `-ch/2`, at the bottom of the canvas in our xy reference frame. Lines 48 and 49 create the two empty lists `hyp1` and `hyp2` that will contain the coordinate samplings for the right and left branch of the hyperbola, respectively. Loop 50–56 fills `hyp1` and `hyp2` by evaluating x as a function of y. It is convenient to evaluate x as a function of y, rather than y a function of x, because $x(y)$ is a double-valued function, while $y(x)$ is a four-valued function. The value of y is increased by 2 px at every step. You are invited to experiment the results of changing the y increase step at Line 56.

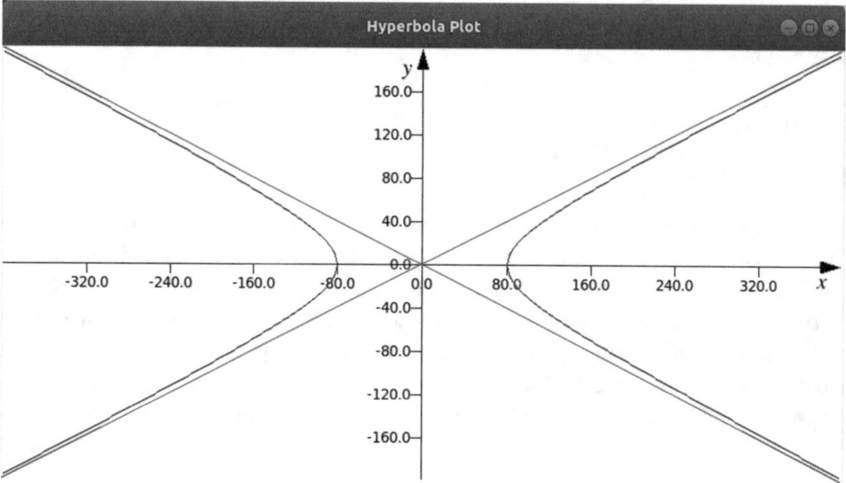

Fig. 6.5 Plotting a hyperbola under Tkinter

```
57 # ............................................ draw hyperbola
58 canvas.create_line(hyp1, fill='blue')
59 canvas.create_line(hyp2, fill='blue')
60 # ..........................................................
61 root.mainloop()
```

Lines 58 and 59 draw the two hyperbola branches.

6.4.2 *Saving an Image of the Canvas*

As we saw in Sect. 3.3, Listing 3.5, the package `matplotlib.pyplot` can save its plots to figures through the method `.savefig()`. Matplotlib can save a figure in the formats *.png*, *.pdf*, *.ps*, *.eps* and *.svg*. Obviously, also the images that we draw on the Tkinter canvas can be saved to figures. For this we have the method `.postscript()`, which saves figures **only** in *Encapsulated PostScript*. This, however, is not a relevant limitation: there are many available programs that can convert image files between different figure formats. The author's (personal) choice is the use of the ImageMagick® package, available for Linux, Windows and macOS, among others. Once ImageMagick® is installed, and you want to convert an Encapsulated PostScript file naned, say, `myfigure.eps` to any other format, say *.jpg*, you simply type

```
convert -density 300 myfigure.eps myfigure.jpg
```

from the command line. Thus, for instance, the last lines of Listing 6.4 must be changed to

```
60 # ..........................................................................
61 canvas.update()
62 canvas.postscript(file='hyperb.eps',colormode='color')
63 root.mainloop()
```

Line 61 updates the canvas, so that the figure can be copied from it. We shall see in Chap. 7 that the method `.update()` is essential in Tkinter animation for refreshing the canvas at each new animation frame. Method `.postscript()` at Line 62 creates the Encapsulated PostScript file `hyperb.eps`. The optional argument `colormode` can have the values `'color'` for color output, `'gray'` for grayscale and `'mono'` for black and white. Since `.postscript()` is a *canvas* mode, only the canvas, not the window frame, is reproduced in the figure file.

6.5 Exercises

Exercise 6.1 Write a script that draws the hexagon of Fig. 6.6

Exercise 6.2 Write a script that plots a Gaussian on the canvas. The equation of the Gaussian is $y = \mathrm{e}^{-x^2/w^2}$, with $w = 2$. Plot the interval $-5 < x < 5$.

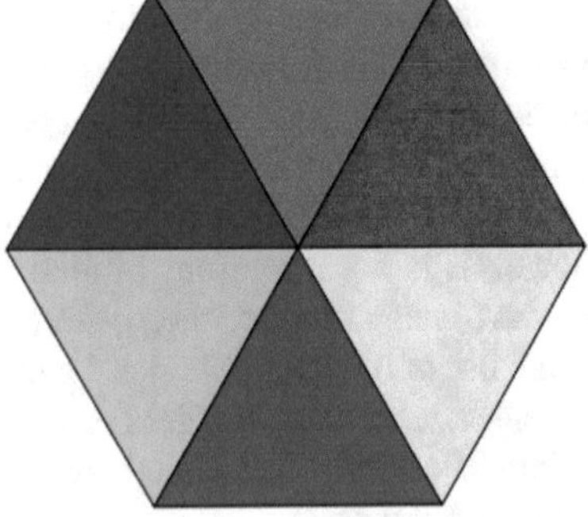

Fig. 6.6 The colors are: red, green, blue, cyan, magenta and yellow

Chapter 7
Tkinter Animation

7.1 Introduction

Tkinter animation is based on the same principle of all computer animations, as well as cinematography and TV: successive images (frames) are shown on the monitor at a given frame rate (24 frames per second in the case of 35 mm sound films), giving the illusion of motion to the human eye. This task is performed by a loop, some times an infinite loop, which, at each cycle,

1. cleans the canvas, removing the preceding frame (from the canvas, not from the monitor!);
2. draws the new frame on the canvas, while the preceding frame is still displayed on the monitor;
3. copies the new frame from the canvas to the monitor;
4. waits an appropriate delay time in order to keep pace with the required frame rate;
5. returns to point 1.

In choosing the frame rate for an *interactive* animation, we must keep in mind that the computer must have the time to perform all calculations needed to draw the new frame, and actually redraw the picture, in the interval between two consecutive pictures. When the calculations involved are too complex, and/or drawing the pictures requires too much time, the animation will be slower than the requested frame rate. In this case, an alternative is asking our program to store all the single frames in a file, which we shall watch as a movie *after* running the program. This, however, will prevent interactivity.

As a simple example, the following code shows a ball (actually, a red circle) moving on the canvas and bouncing at the canvas borders (Fig. 7.1).

Electronic supplementary material The online version of this chapter (https://doi.org/10.1007/978-3-030-45027-4_7) contains supplementary material, which is available to authorized users.

G. Moruzzi, *Essential Python for the Physicist*,
https://doi.org/10.1007/978-3-030-45027-4_7

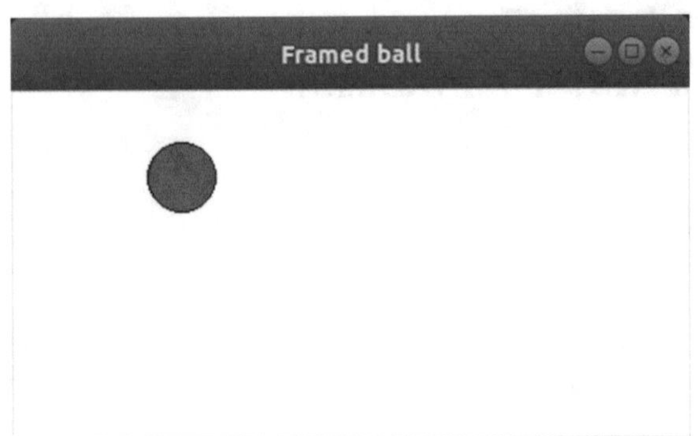

Fig. 7.1 Ball bouncing at the canvas borders

Listing 7.1 FramedBall.py

```
1  #!/usr/bin/env python3
2  from tkinter import Tk,Canvas,ALL
3  # ........................................ open Tkinter root window
4  root=Tk()
5  root.title("Framed ball")
```

Line 4 creates the root window of our program, and Line 5 writes the title of the root window.

```
6  # ........................................ canvas width and height
7  cw=800
8  ch=640
9  # ........................................ add canvas to root window
10 canvas=Canvas(root,width=cw,height=ch,background='white')
11 canvas.grid(row=0,column=0)
12 # ......................................................... variables
13 delay=20 #milliseconds
14 rad=20
15 color="red"
16 x=rad
17 y=ch-rad
18 vx=4.0
19 vy=-5.0
```

Quantities `cw` and `ch` are the width and height of the canvas in pixels, respectively. Line 10 creates the canvas of the required width and height. It also specifies that the canvas belongs to the root window, and that the background color is white. Line 13 sets a delay period of 20 ms, thus, the time interval between two successive frames of our *movie* will be 20 ms *plus* the time needed to calculate and draw a single frame. Lines 14–19 define the variables describing the ball: the radius in px, `rad`, and the color `color`, set to red. Lines 17 and 18 assign the initial x and y coordinates of the ball center: the initial x position equals the radius, while the initial y position equals the canvas height minus the radius. Thus, the ball is located at the lower left corner of the canvas: as usual in computer graphics, the $x = 0$ axis is the canvas left border, while the $y = 0$ axis is the canvas *upper* border, the y axis being directed downwards. Lines 18 and 19 assign the initial x and y components of the ball velocity, `vx` and `vy`, respectively. Actually, `vx` and `vy` are the x and y displacements of the ball at each cycle (each animation step). In other words, we are measuring the ball velocity in "pixels/cycle". Note that a negative y velocity is directed upwards on the monitor.

```
20 # ................................................ main loop
21 while True:
22   # ........................................ draw ball on canvas
23   canvas.delete(ALL)
24   canvas.create_oval(x-rad,y-rad,x+rad,y+rad,fill=color)
25   canvas.update()
26   # ................. is the ball bouncing on the canvas borders?
27   if (x+rad)>=cw:
28     vx=-abs(vx)
29   elif (y+rad)>=ch:
30     vy=-abs(vy)
31   elif x<=rad:
32     vx=abs(vx)
33   elif y<=rad:
34     vy=abs(vy)
35   # ...................... update position and velocity components
36   x+=vx
37   y+=vy
38   # ............................................ wait delay time
39   canvas.after(delay)
```

Lines 21–39 constitute the program's main loop (the *animation loop*). At each cycle, Line 23 clears the canvas, i.e., the preceding frame. Constant `ALL` is imported from Tkinter at Line 2. Line 24 draws the ball in the canvas: an ellipse inscribed in a rectangle (actually, a square, so that the ellipse is a circle) of opposite corners `(x-rad,y-rad)` and `(x+rad,y+rad)`, and filled with the ball color. The circle is thus centered at (x, y). The drawing on the canvas is actually done at Line 25 by the command `canvas.update()`. Lines 27–34 produce the bouncing effect: if the distance between the ball center and one of the canvas borders is equal to or smaller than `rad`, the x or y velocity component is reversed, according to the case. Note that here we are using the function `abs()`, which returns the absolute value of the argument as an integer if the argument is integer, as a float if the argument is a float. Function `math.fabs()` returns the absolute value always as a float, even if the argument is an integer. Lines 36 and 37 update the ball position for the next "movie frame". Line 39 forces the program execution to wait the required delay time between two successive frames. The program is exited by clicking on the red `x` at the upper left corner of the root window. This usually causes the computer to report an error, that you can ignore. When you close the window, your program still tries to do the next iteration, and you get an error message because the canvas no longer exists.

7.2 Adding Uniform Acceleration

Adding a uniform acceleration is very simple: we need to add only two lines to Listing 7.1, and change one line, as discussed in the comments to Listing 7.2. This listing draws a bouncing ball in the presence of gravity (and absence of friction!)

Listing 7.2 GravityBall.py

```
1  #!/usr/bin/env python3
2  from tkinter import Tk,Canvas,ALL
3  # ......................................... open Tkinter root window
4  root=Tk()
5  root.title("Gravity ball")
6  # ........................................... canvas width and height
7  cw=800
8  ch=400
9  # ......................................... add canvas to root window
10 canvas=Canvas(root,width=cw,height=ch,background='white')
11 canvas.grid(row=0,column=1)
12 # ........................................................ variables
13 delay=20 #milliseconds
14 rad=20
15 color="red"
16 x=rad
17 y=ch-rad
18 vx=4.0
19 vy=-7.5
20 ay=0.1
```

Line 20 is the first added line: it assigns the value 0.1 to the variable `ay`, the y (and only nonzero) component of the ball acceleration. The acceleration is measured in "pixels/cycle^2", and, being positive, is directed downwards on the computer monitor, simulating gravity.

```
21 # ........................................................ main loop
22 while True:
23   # ............................................ draw ball on canvas
24   canvas.delete(ALL)
25   canvas.create_oval(x-rad,y-rad,x+rad,y+rad,fill=color)
26   canvas.update()
27   # ................... is the ball bouncing on the canvas borders?
28   if (x+rad)>=cw:
29     vx=-abs(vx)
30   elif (y+rad)>=ch:
31     vy=-abs(vy)
32   elif x<=rad:
33     vx=abs(vx)
34   elif y<=rad:
35     vy=abs(vy)
36   # ....................... update position and velocity components
37   x+=vx
38   y+=vy+0.5*ay
39   vy+=ay
40   # ................................................ wait delay time
41   canvas.after(delay)
```

Line 38 is the line we had to change. The y motion is occurring with uniform acceleration, thus, the y displacement of the ball in a time interval Δt is $\Delta y = v_y\,\Delta t + 0.5\,a_y\,\Delta t^2$. Here we are using the loop *cycle* as time unit, thus $\Delta t = \Delta t^2 = 1$. Line 39 is the last added line, it updates the vertical component of velocity for the next "movie frame".

7.3 Adding Interactive Buttons

Up to now, we could stop and exit our programs only by clicking the mouse on the `x` in the small red circle at the upper left of the window frame. When we do this for Listings 7.1 and 7.2, an error is reported on the terminal because we interrupted the execution of an infinite loop. Another possibility would be to replace the loop command "`while True:`" in the code with, for instance, the command "`for i in range(5000):`", this would stop the program after 5000 cycles. A more refined way is adding *interactive buttons*, which allow us to control the program during its execution. Here follows a listing, where only the additions to, and changes from, Listing 7.2 are commented.

Listing 7.3 ButtonBall.py

```
1  #!/usr/bin/env python3
2  from tkinter import Tk,Button,Canvas,Frame,ALL,W
3  # ............................................ Global variables
4  RunAll=True
5  RunMotion=False
```

Two further functions, `Button()` and `Frame()`, and one further constant, `W`, are imported from Tkinter. Two global variables, `RunAll`, initially set to `True`, and `RunMotion`, initially set to `False`, are added. The program will run as long as `RunAll` is true, the ball will move when `RunMotion` is true, and be in "standby" when `RunMotion` is false. According to Line 5, the ball will stand still at the beginning of the program execution. These two variables are *global*: they are common to the main program and to the functions called by clicking the mouse on the control buttons.

```
6  # ............................................ Start/Stop motion
7  def StartStop():
8    global RunMotion
9    RunMotion=not RunMotion
10   if RunMotion:
11     StartButton["text"]="Stop"
12   else:
13     StartButton["text"]="Restart"
```

Here we define the function `StartStop()`, called by pressing the Start/Stop button, see comments to Lines 30–33 below. This function switches the value of the global variable `RunMotion` from `True` to `False` and vice versa, and changes the label on the Start/Stop button accordingly. If the ball is moving, pressing the button will stop it, and put it in motion if it is in stand-by. The variable `RunMotion` is declared as *global*: it is shared by the main program and `StartStop()`. Failing to declare `RunMotion` as global would cause an error message: Python would interpret the variable as local to `StartStop()`, and complain that Line 9 is using a variable *before* it is assigned a value.

```
14 # ........................................................ Exit program
15 def StopAll():
16   global RunAll
17   RunAll=False
```

Here we define the function `StopAll()`, called by pressing the Close button. This function assigns the value `False` to the global variable `RunAll`, causing the program execution to stop.

```
18 # ................................................ Create root window
19 root=Tk()
20 root.title("Button ball")
21 # ........................................ Add canvas to root window
22 cw=800
23 ch=400
24 canvas=Canvas(root, width=cw, height=ch, background='white')
25 canvas.grid(row=1,column=0)
26 # ....................................... Add toolbar to root window
27 toolbar=Frame(root)
28 toolbar.grid(row=0,column=0,sticky=W)
```

Line 25 locates the canvas at row 1 instead of row 0 as in the previous scripts, because row 0 will be occupied by a horizontal toolbar, containing the control buttons. Line 27 creates the toolbar as a *frame* belonging to the root window, and Line 28 locates it at row 0, column 0, just above the canvas. The option `sticky=W` aligns the buttons at the *left* (*West* in a geographical map) of the toolbar. The other possibility would be `sticky=E` for aligning the buttons at the right. Values `N` and `S` have no effect in a single-row toolbar. Omitting the `sticky` option would center the buttons in the toolbar. Constants `W`, `E`, `N` and `S` are defined by Tkinter.

```
29 # ..................................................... Toolbar buttons
30 StartButton=Button(toolbar,text="Start",command=StartStop)
31 StartButton.grid(row=0,column=0)
32 CloseButton=Button(toolbar, text="Close", command=StopAll)
33 CloseButton.grid(row=0,column=1)
```

Line 30 creates the Start/Stop button, belonging to the toolbar, see Fig. 7.2. The initial text on the button is "Start", because initially the ball is in stand-by. The command associated to the button is the function `StartStop()`, defined at Lines 7–13. Line

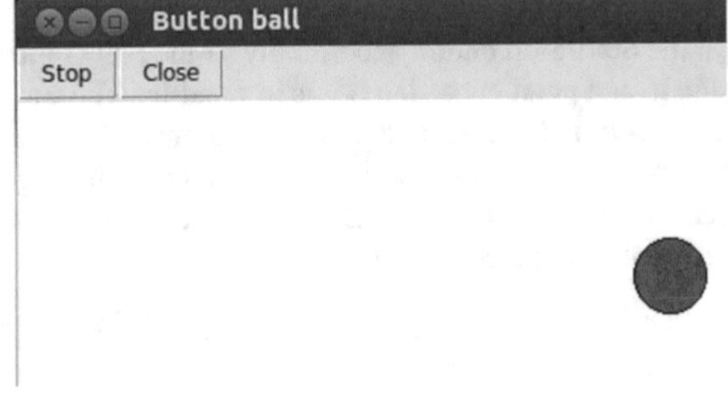

Fig. 7.2 Control buttons on the *toolbar* of the root window

31 locates the button at row 0 (the only row) and column 0 of the toolbar. Lines 32–33 create and locate the Close button, which stops the program execution by calling function `StopAll()`.

```
34 # ........................................................ Variables
35 delay=20 #milliseconds
36 rad=20
37 color="red"
38 x=rad
39 y=ch-rad
40 vx=4.0
41 vy=-7.5
42 ay=0.1
43 # ........................................................ Main loop
44 while RunAll:
45   # ............................................ Draw ball on canvas
46   canvas.delete(ALL)
47   canvas.create_oval(x-rad,y-rad,x+rad,y+rad,fill=color)
48   canvas.update()
49   # ................................. Bouncing on the canvas borders
50   if RunMotion:
51     if (x+rad)>=cw:
52       vx=-abs(vx)
53     elif (y+rad)>=ch:
54       vy=-abs(vy)
55     elif x<=rad:
56       vx=abs(vx)
57     elif y<=rad:
58       vy=abs(vy)
59     # ................................. Update position and velocity
60     x+=vx
61     y+=vy+0.5*ay
62     vy+=ay
63   # ---------------.........................---------- Wait delay time
64   canvas.after(delay)
65   #----------------------------
66 root.destroy()
```

The loop 44–64 iterates as long as `RunAll` is true. When the loop is interrupted, the root window is destroyed at Line 66. During the loop execution, the position of the ball is updated by Lines 49–62 if `RunMotion` is true, otherwise the ball stands still. Now, each time you click the mouse on the Start/stop button of Fig. 7.2 the ball stops if it was moving, and vice versa. If you click on the Close button the program terminates without error messages.

7.4 Numerical Parameters, Entries, Labels and Mouse Dragging

It is often interesting to observe how the program behavior is modified by changing the values of some numerical parameters. A possibility is changing the lines of the listing where the variables are assigned, for instance Line 40 of Listing 7.3 for the initial horizontal velocity `vx`, or Line 42 for the vertical acceleration `ay`. This implies stopping, changing, and restarting the program.

But it is also possible to change the parameter values interactively, during the program execution. For this, we need to introduce appropriate *labels* and *entries* in the toolbar, as shown in Fig. 7.3. After clicking the mouse on the entry widget, the user can enter a new value by typing it on the keyboard. Then, the program can read the new text on the entry and perform the required conversions. The label widget is a standard Tkinter widget used to display a text or image on the screen. Here we shall use it to specify the name of the parameter displayed in the nearby entry.

Further, it is possible to change the initial conditions by dragging the objects on the canvas by means of the mouse. Listing 7.4 follows, as usual only the changes from Listing 7.3 are commented.

Listing 7.4 MouseEntryBall.py

```
#!/usr/bin/env python3
from tkinter import *
# ............................................... Global variables
RunAll=True
RunMotion=GetData=Grabbed=False
# ............................................... Start/Stop motion
def StartStop():
  global RunMotion
  RunMotion=not RunMotion
  if RunMotion:
    StartButton["text"]="Stop"
  else:
    StartButton["text"]="Restart"
# ................................................... Exit program
def StopAll():
```

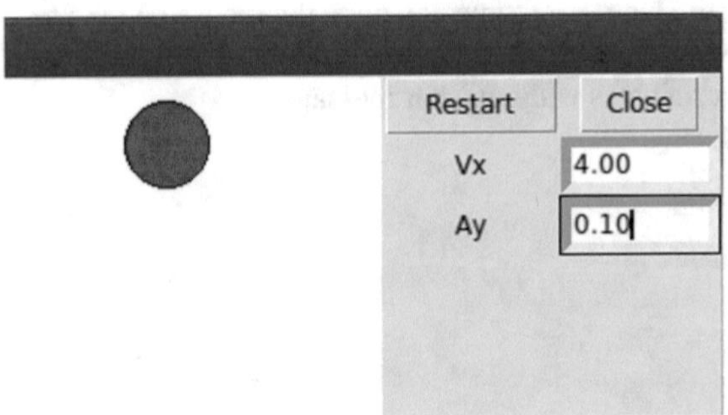

Fig. 7.3 Control buttons, *labels* and *entries* on the toolbar

```
16     global RunAll
17     RunAll=False
18  # ................................................... Read entries
19  def ReadData(*arg):
20     global GetData
21     GetData=True
```

From this listing on, we are importing *everything* from Tkinter at Line 2, in spite of this being discouraged at the end of Sect. 1.8. From the author's experience, importing everything from the Tkinter package (and from Tkinter only!) never led to variable conflicts. You are simply advised to be careful.

The new global variables `GetData` and `Grabbed`, both initially set to *False*, are added at Line 5. When the motion is in standby, if `Getdata` is *True* the program will read new values from the entries, if `Grabbed` is *True* the mouse cursor will grab the ball and move it on the canvas. The value of `GetData` is turned to `True` by the new function `ReadData()`, defined at Lines 18–21.

```
22  # ..................................................... Grab ball ****
23  def GrabBall(event):
24     global Grabbed,rad,RunMotion,x,y
25     if not RunMotion:
26        Grabbed=((x-event.x)**2+(y-event.y)**2)<rad**2
27  # ................................................... Release ball ****
28  def ReleaseBall(event):
29     global Grabbed
30     Grabbed=False
31  # ..................................................... Drag ball ***
32  def DragBall(event):
33     global Grabbed,x,y
34     if Grabbed:
35        x,y=event.x,event.y
```

All three functions above have a single argument, `event`. Python *events* correspond to various actions, for instance the pressing of a keyboard key, or a mouse operation. These are the only events we are interested in here.

If the ball is in standby (`RunMotion` is *False*), function `GrabBall()` checks if the distance of the mouse pointer, located on the canvas at `(event.x, event.y)`, from the ball center, located at `(x,y)`, is smaller than the ball radius, `rad`. In this case the ball is grabbed by the mouse cursor by setting `Grabbed` equal to *True*. Pressing the mouse left button is the *event* that activates function `GrabBall()`, see Line 46 below.

The event that calls function `ReleaseBall()`, defined at Lines 27–30, is releasing the left mouse button, see Line 48 below. The function releases (frees) the grabbed ball, and sets `Grabbed` to *False*.

Function `DragBall()`, defined at Lines 31–35, copies the mouse-cursor coordinates `(event.x, event.y)` into the ball-center coordinates `(x,y)`, thus forcing the ball to follow the mouse-cursor movements on the canvas. The event activating this function, provided that the global variable `Grabbed` is *True*, is the motion of the mouse cursor over the canvas, see Line 47 below.

```
36  # ............................................. Create root window
37  root=Tk()
38  root.title("Entry ball")
39  root.bind('<Return>',ReadData)
40  # ....................................... Add canvas to root window
41  cw=800
42  ch=400
43  canvas=Canvas(root, width=cw, height=ch, background='white')
44  canvas.grid(row=0,column=0)
45  # ...................................................... Mouse button ***
46  canvas.bind('<Button-1>',GrabBall)
47  canvas.bind('<B1-Motion>',DragBall)
48  canvas.bind('<ButtonRelease-1>',ReleaseBall)
```

Line 39 binds the *Return* key of the keyboard to the function ReadData(): whenever the *Return* key is pressed, ReadData() is called, and GetData is set to *True*. See Appendix F.

Line 46 binds the event "pressing the left mouse button" to the function GrabBall(), defined at Lines 23–26, provided that the mouse cursor is inside the canvas. In Tkinter notation, event <Button-1> corresponds to pressing the left mouse button, <Button-2> to pressing the middle mouse button, and <Button-3> to pressing the right button. Codes <Button-4> <Button-5> refer to the events of turning the mouse wheel forward and backward. Again, see Appendix F.

Line 47 binds the event "moving the mouse while the left button is pressed" to the function DragBall(). Codes <B2-motion> and <B3-motion> refer to mouse motion while the middle, or right, button is pressed, respectively.

Line 48 binds the event "releasing the left mouse button" to the function ReleaseBall(). Thus, pressing the left mouse button when the mouse cursor is on the ball (closer to the ball center than the ball radius) will grab the ball. Moving the mouse keeping the left button pressed will drag the ball over the canvas, and releasing the left mouse button (code <ButtonRelease-1>) will free the ball.

```
49  # ...................................... Add toolbar to root window
50  toolbar=Frame(root)
51  toolbar.grid(row=0,column=1,sticky=N)
52  # ............................................... Toolbar buttons
53  StartButton=Button(toolbar,text="Start",command=StartStop,width=7)
54  StartButton.grid(row=0,column=0)
55  CloseButton=Button(toolbar, text="Close", command=StopAll)
56  CloseButton.grid(row=0,column=1)
```

The Start button and the Close button are located in *subcolumns* 0 and 1 of the toolbar, respectively. In this program we use a vertical toolbar, located in column 1, at the right of the canvas (located in column 0), see Fig. 7.3.

```
57  # ..................................... Toolbar labels and entries
58  LabVx=Label(toolbar,text="Vx")
59  LabVx.grid(row=1,column=0)
60  EntryVx=Entry(toolbar,bd=5,width=8)
61  EntryVx.grid(row=1,column=1)
```

```
62 LabAccel=Label(toolbar,text="Ay")
63 LabAccel.grid(row=2,column=0)
64 EntryAccel=Entry(toolbar,bd=5,width=8)
65 EntryAccel.grid(row=2,column=1)
```

Lines 58 and 59 create the label `LabVx` and locate it in subrow 1, column 0 of the toolbar: subrow 0 is occupied by the buttons. Lines 60 and 61 create the entry where we can type new values for the horizontal velocity component, and locate it at the right of the corresponding label, at subrow 1 and subcolumn 1 of the toolbar. Line 60 also specifies that the border of the entry is 5 pixels (`bd=5`), and that the entry will show a maximum of 8 characters (`width=8`). Lines 62-65 create the label and the entry for the vertical acceleration `ay`.

```
66  # ......................................................... Variables
67  delay=20 #milliseconds
68  rad=20
69  color="red"
70  x=rad
71  y=ch-rad
72  vx=4.0
73  vy=-7.5
74  ay=0.1
75  # ............................... Write variable values into entries
76  EntryVx.insert(0,'{:.2f}'.format(vx))
77  EntryAccel.insert(0,'{:.2f}'.format(ay))
78  # ......................................................... Main loop
79  while RunAll:
80    # ............................................ Draw ball on canvas
81    canvas.delete(ALL)
82    canvas.create_oval(x-rad,y-rad,x+rad,y+rad,fill=color)
83    canvas.update()
84    # ................................................. Ball is moving
85    if RunMotion:
86      # ...................................................... Bouncing
87      if (x+rad)>=cw:
88        vx=-abs(vx)
89      elif (y+rad)>=ch:
90        vy=-abs(vy)
91      elif x<=rad:
92        vx=abs(vx)
93      elif y<=rad:
94        vy=abs(vy)
95      # ................................ Update position and velocity
96      x+=vx
97      y+=vy+0.5*ay
98      vy+=ay
99    # .................................................. Read entries
100   elif GetData:
101     try:
102       vx=float(EntryVx.get())
103     except ValueError:
104       pass
105     try:
```

```
106           ay=float(EntryAccel.get())
107         except ValuError:
108           pass
109         EntryVx.delete(0,'end')
110         EntryVx.insert(0,'{:.2f}'.format(vx))
111         EntryAccel.delete(0,'end')
112         EntryAccel.insert(0,'{:.2f}'.format(ay))
113         GetData=False
114     # ........................................ Wait delay time
115     canvas.after(delay)
116     #——————————————————————
117   root.destroy()
```

Lines 100–113 are effective only when the ball is in standby (when `RunMotion` is *False*) and `GetData` is *True*. Line 102 reads what is typed in the entry `EntryVx` and converts it to a float value, assigned to the variable `vx`. The reason for the `try` statement at Line 101 is that you might have typed some characters that are not numbers in the entry. In this case the function `float()` could not convert the value and would report an *exception* (in Python, errors detected during program execution are called *exceptions*). Lines 101–104 tell the code to try if it is possible to convert the entry string into a float value, if the conversion is successful, the value is assigned to `vx`. If a `ValueError` exception is raised, i.e., the conversion was not successful, Line 104 tells the program to do nothing: `vx` preserves its old value. Lines 105-108 check if the string typed into the entry `EntryAccel` can be converted into a new numerical value for `ay`. Line 109 deletes the present content of the entry `EntryVx` from 0, i.e., the first character of the string, to the end of the string. Line 110 rewrites the new value formatted with two digits after the decimal point. Line 113 resets the variable `GetData` to *False*.

You can type new values for the horizontal velocity and for the vertical acceleration of the ball at any time, provided that the ball is in standby. The corresponding variables will assume the new values only after you press both the *Return* key and the *Restart* button, no matter in which order. When the ball is in standby, you can also grab it and change its position with the mouse.

7.5 Creating Video Files Under Tkinter

It is often convenient to create video files from our Python animations, both for sending them to friends (or for showing them in conferences), and because animations can involve complex computations between successive frames, causing the program to run very slowly. Unfortunately Tkinter does not have any method equivalent to `.FuncAnimation()` that we met in Sect. 3.7. The easiest way is using one of the many available programs that record the entire computer screen or a selected part of it.

A good choice for Linux is `SimpleScreenRecorder` by Maarten Baert. It is easy to install and to use, and has the possibility to reduce the video frame rate if your animation is running too slow due to the complex computations mentioned above. Recording the screen of a Mac is easy if you are using macOS Mojave: just hit the keyboard shortcut Shift+cmd+5 and all the controls for capturing video and still images from your desktop will appear. You can record the whole screen, a section, or a specific window, then trim, save or share the resulting footage. Windows 10 offers the built-in Xbox app, featuring screen capturing tools. Launch the Xbox app, then press the Windows and G icons on the keyboard and choose 'Yes this is a game' option. If you want to change the video quality or adjust the audio settings, you can do so by opening Game DVR options menu. If you are satisfied with one of the above options, or if you have found another screen-capturing application of your taste, you can skip the rest of Sect. 7.5. Honestly, you are advised to do so.

Creating a video file under Tkinter is possible and absolutely not complicated. However, it is true that it can be somewhat time-consuming, but it is the computer's time, not necessarily yours! The basic idea is copying each animation frame into a picture file, then merging all the pictures into a single video file with the help of an external program. As we saw in Sect. 6.4.2, the method `canvas.postscript()` saves the canvas content into an Encapsulated PostScript (*eps*) file. Thus, if we want to create a video file from Script 7.4, we start by modifying Lines 78–94 as follows

Listing 7.5 EntryBall.py Modified to Create Video Files

```
 75 # .............................. Write variable values into entries
 76 EntryVx.insert(0,'{:.2f}'.format(vx))
 77 EntryAccel.insert(0,'{:.2f}'.format(ay))
 78 # ................................................. Frame counter
 79 iFrame=0
 80 # ...................................................... Main loop
 81 while RunAll:
 82   # ........................................... Draw ball on canvas
 83   canvas.delete(ALL)
 84   canvas.create_oval(x-rad,y-rad,x+rad,y+rad,fill=color)
 85   canvas.update()
 86   # ................................................ Ball is moving
 87   if RunMotion:
 88     # ......................................... Create video frame
 89     FrameName='../VideoFrames/frame{:08d}.eps'.format(iFrame)
 90     canvas.postscript(file=FrameName,colormode='color')
 91     iFrame+=1
 92     # .................................................. Bouncing
 93     if (x+rad)>=cw:
 94       vx=-abs(vx)
 95     elif (y+rad)>=ch:
 96       vy=-abs(vy)
 97     elif x<=rad:
 98       vx=abs(vx)
 99     elif y<=rad:
100       vy=abs(vy)
```

where we have added Lines 79 and 88–91. Variable `iFrame`, defined at Line 79, is an animation-frame counter. Lines 88–91 are executed only if `RunMotion` is *True* (if the ball is moving). Line 89 creates a name for the *eps* frame picture in the form `path/frameXXXXXXXX.eps`, where `XXXXXXXX` is a zero-padded integer number ranging from 0000000000 to 99999999, equal to the frame counter. Actually, in all practical cases, the highest number will be much smaller, since 10^8 is a very high number of frames. Even running at 50 frames/second, a video comprising 10^8 frames would last $10^8/50 = 2 \times 10^6$ s $\simeq$ 555.5 hours! For the path, here we are assuming that you have previously created an empty directory named `VideoFrames` parallel to your working directory, where the animation frames will be stored. Under Windows all slashes (/) in the path must be replaced by backslashes (\). Line 90 copies the current animation-frame canvas into the *eps* file, and Line 91 increases the frame counter.

When we stop our modified program the `VideoFrames` directory will contain all our animation frames in *eps* format. Unfortunately there is no program that can directly merge *eps* files into a video file. Thus, we must first convert our *eps* files to some other format. Here we shall convert them to *jpg* (Joint Photographic Experts Group), but also several other formats, like, for instance, *png* (Portable Network Graphics), would do the job. The program `convert` of the ImageMagick® package can do the conversion for us. Our *jpg* files can then be merged into a video file, for instance *mp4* (MPEG-4: Moving Picture Experts Group), or *avi* (Audio Video Interleave, created by Microsoft), by the program `ffmpeg`. *FFmpeg* is a free software project consisting of a vast software suite of libraries and programs for handling video, audio and other multimedia files and streams. It is available for the Linux, Windows and macOS platforms. Listing 7.6 does the whole conversion from the original *eps* files to the final video file. All you have to do is typing, for instance,

`MakeVideo.py bouncing.mp4` or `MakeVideo.py bouncing.avi` in the `VideoFrames` directory, and the script will create a video file for you.

Listing 7.6 MakeVideo.py

```
#!/usr/bin/env python3
import time
import os
from sys import argv
# ................................................ Convert frames
movie=argv[1]
tt0=time.time()
nn=0
for InFile in os.listdir('.'):
    if InFile.endswith('.eps'):
        base=os.path.splitext(InFile)[0]
        OutFile=base+".jpg"
        command="convert -density 300 ./"+InFile+" -flatten ./"+OutFile
        os.system(command)
        nn+=1
ttt=time.time()
print(ttt-tt0,' seconds')
print((ttt-tt0)/nn,' seconds/frame')
# ...................................... Merge frames to video file
tt0=ttt
command='ffmpeg -r 50 -f image2 -i ./frame%08d.jpg -vcodec libx264'\
```

```
22        +' -crf 25 -vf scale=1280:-2 -pix_fmt yuv420p '+movie
23    os.system(command)
24    ttt=time.time()
25    print(ttt-tt0,' seconds for FFmpeg')
```

Line 6 copies the second command-line argument, in our case `bouncing.mp4` or `bouncing.avi` according to your choice, into the string variable `movie`, see the discussion of Listing 2.1. Line 7 stores the initial time of the script execution into `tt0`, this will be needed for evaluating the computation time. Variable `nn` at Line 8 is an animation-frame counter, needed at line 18 for evaluating the conversion time per animation frame. Loop 9–15 converts our *eps* files to *jpg*. At Line 9, the method `os.listdir('.')` returns a list comprising the entries of the current directory, specified by the path `'.'`. Thus, the variable `InFile` iterates over all the directory entries. The string method `.endswith()` at Line 10 returns *True* if the string `InFile` ends with the suffix `.eps`, otherwise returns *False*. Thus, Lines 11–15 are executed only for *eps* files.

Method `os.path.splitext()` at Line 11 splits its argument, `InFile`, into a string pair (`root`, `ext`) such that `root` + `ext` = `InFile`, and `ext` is either empty or begins with a period and contains at most one period. Thus, if `InFile` is, for instance, `frame00000015.eps`, the string `root` is `frame00000015`, and `ext` is `.eps`. As a result, `base` is `frame00000015`, and the string `OutFile` is `frame00000015.jpg`. Line 13 builds the command string to be passed as argument to the method `os.system`, see Sect. 2.4. Line 14 calls the external command

```
convert -density 300 frameXX.eps -flatten frameXX.jpg
```

that converts the *eps* file to a *jpg* file. The `XX` in the file names stands for an 8-digit integer (the frame counter) with the appropriate number of leading zeros. The command option `-density` specifies the image resolution to store when converting a vector image, like *eps* in our case, to a *raster graphics*, also called *bitmap*, image such as *jpg*, *pnm* or *png*. The default resolution is 72 dots per inch, which is equivalent to one pixel per typographic point (1 point = 1/72 in.). A value of 300, as chosen here, will lead to satisfactory results for all practical purposes. The command option `-flatten` is needed to preserve our background color, otherwise some output formats, like *png*, might have a transparent background. Line 15 increases by one the frame counter.

Line 16 gets the time after all frames have been converted, and Lines 17–18 print the time needed for the whole conversion process, and the average time per single-image conversion, on the terminal. The time needed for converting a single image is of the order of 1 s, ranging from some 0.5 s up to some 3 s depending on the image complexity and on the computer speed. At a frame rate of 24 fps (frames per second), an animation of 10 min comprises 14,400 frames, whose conversion from *eps* to *jpg* thus requires a time of the order of 4 h. But, once you have launched the conversion script, you can let your computer do the work alone overnight, if you wish. If you have a multiprocessor computer you can divide the conversion time by approximately a factor n, where n is the number of processors.

Lines 21–22 do the final job, merging the converted *jpg* frames into the output mp4, or *avi*, video file, by calling the external command ffmpeg. The external command is

```
ffmpeg -r 24 -f image2 -i ./frame\%08d.jpg -vcodec \
libx264 -crf 25 -vf scale=1280:-2 -pix\_fmt yuv420p movie
```

where movie is either bouncing.mp4 or bouncing.avi. These are the meanings of the command options:

- - r 24 stands for a frame rate of 24 fps, you can experiment with different values, obtaining effects from time-stretching (digital slow motion) to fast motion. An extremely low frame rate, like -r 1/5 gives each image a duration of 5 seconds, good for a *slide show*, not for an animation. An extremely high frame rate can exceed the computer capabilities.
- -f image2 tells *FFmpeg* that the input is a sequence of separate images, to be merged into a single video file.
- -i ./frame%08d.png specifies that the input files are in the current directory (./), and their names have the form framexxxxxxxx.png, where xxxxxxxx is a progressive 8-digit integer patched with trailing zeros.
- -vcodec libx264 specifies the computer program used for en**cod**ing and **dec**oding the digital data stream, here libx264, which is free software available for Linux, Windows and macOS. Usually libx264 is installed automatically when you install *FFmpeg*.
- -crf 25 sets the quality/size tradeoff for constant-quality (no bitrate target) and constrained-quality (with maximum bitrate target) modes. Valid range is 0 to 63, higher numbers indicating lower quality and smaller output size. A value of 25 is a reasonable compromise, while 63 leads to a very low quality. The default value is 23.
- -vfscale=1280:-2 is a scale filter which resizes the image to a horizontal width of 1280 pixels. You can specify both width and height by typing, for instance, -vfscale=1280:800, thus changing the aspect ratio. If you want to keep the aspect ratio you can type -1 for the vertical size, -vfscale=1280:-1, this will calculate the height of the output image according to the aspect ratio of the input image. Some codecs require the size of width and height to be a multiple of a certain number n. You can achieve this by setting the width or height to $-n$, as in the present case.
- -pix_fmt yuv420p ensures compatibility with a wide range of playback programs. It is required here, for example, for the video to be playable by Windows Media Player
- movie, has been set equal to our second command-line argument at Line 6, i.e., either bouncing.mp4 or bouncing.avi, according to your choice. It is the name of the video output file. If *mp4* or *avi* are not convenient for you, FFmpeg supports many common and some uncommon image formats, like, for instance, *gif*.

If you wish, you can change the file extensions .jpg at Lines 12 and 21 with the extensions .png, .pnm or other. The program will run anyway, in our experiments we found that the conversion to `.jpg` was faster, but you might want to experiment a little yourself.

As stated above, in a *multiprocessor*, or *multicore*, computer, the time needed for converting the animation frames from *eps* to *jpg* can be reduced by a factor approximately equal to the number of "logical processors". This is done by letting the logical processors work in parallel as much as possible. Script 7.7 is an example of how we can do this. Assuming that you have 4 logical processors, the usage is

```
MakeVideoParallel.py 4 bouncing.mp4
```

where the number 4 can be replaced by the actual number of available processors, and the extension `.mp4` can be replaced by the extension `.avi` or whatever video format you prefer.

Listing 7.7 MakeVideoParallel.py

```
#!/usr/bin/env python3
import math
import time
import os
from sys import argv,exit
# ...................................... Is script called correctly?
if len(argv)!=3:
  print('usage: MakeVideoParallel.py NumberOfProcessors  OutputFile')
  exit()
# ..................... Number of processors and name of output file
nproc=int(argv[1])
movie=argv[2]
# ............................. Form lists of input and output files
ListIn=[]
ListOut=[]
for InFile in os.listdir("."):
  if InFile.endswith(".eps"):
    ListIn.append(InFile)
    base=os.path.splitext(InFile)[0]
    OutFile=base+".jpg"
    ListOut.append(OutFile)
nn=len(ListIn)
# ......................... Convert animation frames from eps to jpg
tt0=time.time()
i=0
while i<nn:
  j=0
  command=''
  while j<nproc:
    if (i+j)<nn:
      command+="convert -density 300 ./"+ListIn[i+j]+" -flatten ./"\
        +ListOut[i+j]
    if j<(nproc-1) and (i+j)<(nn-1):
      command+=" | "
    j+=1
  os.system(command)
  i+=nproc
ttt=time.time()
```

```
39 print(ttt-tt0,' seconds')
40 print((ttt-tt0)/nn,' seconds/image')
41 # .................................. Merge jpg files into video file
42 tt0=ttt
43 command='ffmpeg -r 24 -f image2 -i ./frame%08d.jpg -vcodec libx264'\
44   +' -crf 25 -vf scale=1280:-2 -pix_fmt yuv420p '+movie
45 os.system(command)
46 ttt=time.time()
47 print(ttt-tt0,'seconds for FFmpeg')
```

Lines 7–9 check if the number of entered command-line arguments is correct, they must be 3: (i) `MakeVideoParallel.py`, (ii) the number of processors, (iii) the name of the output video file. If this is not the case the script warns you and exits. Line 11 copies the number of processors to be used into the integer variable `nproc`, Line 12 copies the name of the output video file into the string variable `movie`.

Lines 14–22 form the lists of the names of the input *eps* files, `ListIn`, and of the output *jpg* files, `ListOut`. The integer variable `nn` is the number of animation frames to be converted.

Lines 24–40 perform the format conversion. They build composit command strings of the type

```
convert -density 300 ./frame00000000.eps -flatten frame0000000.jpg |
convert -density 300 ./frame00000001.eps -flatten frame0000001.jpg |
...
```

where the vertical bars " | " are used to join separate commands into a single command string, to be passed as argument to the method `os.system()`. Thus, Line 36 launches the joined commands simultaneously, and they can be simultaneously executed by different processors, if available. The rest of the script is analogous to Listing 7.6.

7.6 Animation and Ordinary Differential Equations

7.6.1 Euler's Method

A motion occurring with uniform acceleration is just a special case. In the general case the force acting on a body will be some function of both the body position and velocity, and, because of Newton's second law, we must deal with differential equations. In most cases the differential equations describing the motion have no analytical solution, and we must use the numerical methods discussed in Chap. 5. As a first simple example, simple because the acceleration is actually constant, we shall apply Euler's method, presented in Sect. 5.2, to the program of our Listing 7.4. As usual, we shall comment only the changes to the original listing.

Listing 7.8 GravityEuler.py

```
#!/usr/bin/env python3
```

```
2  from tkinter import *
3  # ............................................... Global variables
4  RunAll=True
5  RunMotion=GetData=False
6  # ............................................... Start/Stop motion
7  def StartStop():
8    global RunMotion
9    RunMotion=not RunMotion
10   if RunMotion:
11     StartButton["text"]="Stop"
12   else:
13     StartButton["text"]="Restart"
14 # ............................................... Exit program
15 def StopAll():
16   global RunAll
17   RunAll=False
18 # ............................................... Read entries
19 def ReadData(*arg):
20   global GetData
21   GetData=True
22 # ............................................... Variables
23 delay=20 #milliseconds
24 rad=20
25 color="red"
26 x=rad
27 y=rad
28 vx=4.0
29 vy=7.5
30 ay=-0.2
31 mass=10
32 ener=mass*(0.5*(vx**2+vy**2)-ay*y)
```

Two new variables are added: the mass of the ball, set equal to 10 in arbitrary units, and the energy of the ball, evaluated at Line 32 as the sum of the potential and kinetic energies.

```
33 # ............................................... Create root window
34 root=Tk()
35 root.title("Gravity Euler")
36 root.bind('<Return>',ReadData)
37 # ............................................... Add canvas to root window
38 cw=800
39 ch=400
40 canvas=Canvas(root, width=cw, height=ch, background='white')
41 canvas.grid(row=0,column=0)
42 # ............................................... Add toolbar to root window
43 toolbar=Frame(root)
44 toolbar.grid(row=0,column=1,sticky=N)
45 # ............................................... Toolbar buttons
46 nr=0
47 StartButton=Button(toolbar,text="Start",command=StartStop,width=7)
48 StartButton.grid(row=nr,column=0)
49 CloseButton=Button(toolbar, text="Close", command=StopAll)
50 CloseButton.grid(row=nr,column=1)
```

```
51 nr+=1
52 # ...................................... Toolbar labels and entries
53 LabVx=Label(toolbar,text="Vx")
54 LabVx.grid(row=nr,column=0)
55 EntryVx=Entry(toolbar,bd=5,width=8)
56 EntryVx.grid(row=nr,column=1)
57 nr+=1
58 LabAccel=Label(toolbar,text="Ay")
59 LabAccel.grid(row=nr,column=0)
60 EntryAccel=Entry(toolbar,bd=5,width=8)
61 EntryAccel.grid(row=nr,column=1)
62 nr+=1
63 # .................................................... Energy label
64 EnerLab0=Label(toolbar,text='Energy:',font=("Helvetica",11))
65 EnerLab0.grid(row=nr,column=0)
66 EnerLab=Label(toolbar,text='{:8.3f}'.format(ener),font=("Helvetica",11))
67 EnerLab.grid(row=nr,column=1,sticky=W)
68 nr+=1
69 # ............................... Write variable values into entries
70 EntryVx.insert(0,'{:.2f}'.format(vx))
71 EntryAccel.insert(0,'{:.2f}'.format(ay))
```

Two new labels are added to the toolbar at lines 63–67. Label `EnerLab0`, defined at Lines 64–65. simply contains the word “Energy”, while Label `EnerLab`, defined at Lines 66–67 contains the value of the total energy of the bouncing ball. Its value will be updated every ten iterations of the main animation loop.

```
72 # ........................................................ Main loop
73 count=0
74 while RunAll:
75   # .............................................. Draw ball on canvas
76   canvas.delete(ALL)
77   canvas.create_oval(x-rad,ch-(y+rad),x+rad,ch-(y-rad),fill=color)
78   canvas.update()
79   # ................................................ Ball is moving
80   if RunMotion:
81     # .................................................... Bouncing
82     if (x+rad)>=cw:
83       vx=-abs(vx)
84     elif (y+rad)>=ch:
85       vy=-abs(vy)
86     elif x<=rad:
87       vx=abs(vx)
88     elif y<=rad:
89       vy=abs(vy)
90     # ................ Update position and velocity, Euler algorithm
91     x+=vx
92     y+=vy
93     vy+=ay
```

The new variable `count`, initialized to zero, is defined at line 73. Lines 74–116 constitute the main animation loop of the program. The Euler method is a first-order method, with velocity and position updated at each iteration at Lines 91–93. The

x component of velocity is constant between consecutive bouncings at the left and right borders of the canvas. The differential equation for the y motion is

$$m \frac{d^2y}{dt^2} = ma_y \,, \tag{7.1}$$

which, according to Sect. 5.1, can be rewritten as the system of two first-order differential equations

$$\begin{aligned} \frac{dy}{dt} &= v_y \\ \frac{dv_y}{dt} &= a_y \,, \end{aligned} \tag{7.2}$$

which Euler's method solves by the recursive formulas (5.5)

$$y_{i+1} = y_i + v_i \, \Delta t \,, \quad v_{y,i+1} = v_{y,i} + a_y \, \Delta t \,, \tag{7.3}$$

corresponding to Lines 92 and 93, since we have $\Delta t = 1$ in our units.

```
 94      # ............................................ Read entries
 95      elif GetData:
 96        try:
 97          vx=float(EntryVx.get())
 98        except ValueError:
 99          pass
100        try:
101          ay=float(EntryAccel.get())
102        except ValuError:
103          pass
104        EntryVx.delete(0,'end')
105        EntryVx.insert(0,'{:.2f}'.format(vx))
106        EntryAccel.delete(0,'end')
107        EntryAccel.insert(0,'{:.2f}'.format(ay))
108        GetData=False
109      # ............................................ Write energy
110      count+=1
111      if count >=10:
112        count=0
113        ener=mass*(0.5*(vx**2+vy**2)-ay*y)
114        EnerLab['text']='{:8.3f}'.format(ener)
115      # ......................................... Wait delay time
116      canvas.after(delay)
117      #——————————————————
118  root.destroy()
```

The variable `count` is incremented at Line 110. Every ten lines the total energy of the bouncing ball is evaluated at line 113 and displayed on the toolbar label.

Euler integration is a first-order method, and its limits are apparent if we run Script 7.8 for a few minutes: the ball jumps slowly become higher and higher, implying that energy is not conserved. This is seen also by looking at the value displayed by the energy label in the toolbar. The reason is simple: the first of (7.3) assumes a constant velocity during the execution of each step (obviously, the velocity is different from step to step), equal to the initial velocity of the step. But, in reality, the velocity decreases during the step because a_y is negative. Thus, independently of the ball going upwards or downwards, the calculated position at the end of each step is slightly higher than the correct position, leading to a slow increase in the calculated energy. Since a_y is constant, all errors have the same sign and there is no hope of random cancellations.

7.6.2 The Leapfrog Method

In contrast to Euler integration, leapfrog integration is a second-order method, yet it requires the same number of function evaluations per step. Unlike Euler integration, it is stable for oscillatory motion, as long as the time-step Δt is constant and $\Delta t \leq 1/\omega$, ω being the angular frequency of the oscillation. In leapfrog integration, the equations for updating position and velocity are

$$\begin{aligned} x_i &= x_{i-1} + v_{i-\frac{1}{2}}\,\Delta t\ , \\ a_i &= F(x_i)\ , \\ v_{i+\frac{1}{2}} &= v_{i-\frac{1}{2}} + a_i\,\Delta t\ , \end{aligned} \tag{7.4}$$

with positions and accelerations evaluated at "integer times" $i\,\Delta t$, and velocities evaluated at "half-integer" times $\left(i+\frac{1}{2}\right)\Delta t$. The advantage of the method is that, although the position is still updated assuming a constant velocity at each step, the velocity is calculated at the middle of the step rather than at one of the end points. Equation (7.4) can be re-arranged to the "kick-drift-kick" form

$$\begin{aligned} v_{i+\frac{1}{2}} &= v_i + a_i\,\frac{\Delta t}{2}\ , \\ x_{i+1} &= x_i + v_{i+\frac{1}{2}}\,\Delta t\ , \\ v_{i+1} &= v_{i+\frac{1}{2}} + a_{i+1}\,\frac{\Delta t}{2}\ , \end{aligned} \tag{7.5}$$

used in Script 7.9, where Euler integration of Listing 7.8 is replaced by leapfrog integration. Note that leapfrog integration cannot be applied in the presence of a velocity-dependent accelerations.

Listing 7.9 GravityFrog.py

```
#!/usr/bin/env python3
from tkinter import *
# ............................................ Global variables
RunAll=True
```

```
RunMotion=GetData=False
# ............................................ Start/Stop motion
def StartStop():
  global RunMotion
  RunMotion=not RunMotion
  if RunMotion:
    StartButton["text"]="Stop"
  else:
    StartButton["text"]="Restart"
# ................................................... Exit program
def StopAll():
  global RunAll
  RunAll=False
# ................................................... Read entries
def ReadData(*arg):
  global GetData
  GetData=True
# ...................................................... Variables
delay=20 #milliseconds
rad=20
color="red"
x=rad
y=rad
vx=4.0
vy=7.5
ay=-0.1
mass=10
ener=mass*(0.5*(vx**2+vy**2)-ay*y)
# ............................................ Create root window
root=Tk()
root.title('Gravity Leapfrog')
root.bind('<Return>',ReadData)
# .................................... Add canvas to root window
cw=800
ch=400
canvas=Canvas(root, width=cw, height=ch, background='white')
canvas.grid(row=0,column=0)
# ................................... Add toolbar to root window
toolbar=Frame(root)
toolbar.grid(row=0,column=1,sticky=N)
# ............................................. Toolbar buttons
nr=0
StartButton=Button(toolbar,text="Start",command=StartStop,width=7)
StartButton.grid(row=nr,column=0)
CloseButton=Button(toolbar, text="Close", command=StopAll)
CloseButton.grid(row=nr,column=1)
nr+=1
# .................................. Toolbar labels and entries
LabVx=Label(toolbar,text="Vx")
LabVx.grid(row=nr,column=0)
EntryVx=Entry(toolbar,bd=5,width=8)
EntryVx.grid(row=nr,column=1)
nr+=1
```

```
58 LabAccel=Label(toolbar,text="Ay")
59 LabAccel.grid(row=nr,column=0)
60 EntryAccel=Entry(toolbar,bd=5,width=8)
61 EntryAccel.grid(row=nr,column=1)
62 nr+=1
63 # ................................................. Energy label
64 EnerLab0=Label(toolbar,text='Energy:',font=("Helvetica",11))
65 EnerLab0.grid(row=nr,column=0)
66 EnerLab=Label(toolbar,text='{:8.3f}'.format(ener),font=("Helvetica",11))
67 EnerLab.grid(row=nr,column=1,sticky=W)
68 # ............................... Write variable values into entries
69 EntryVx.insert(0,'{:.2f}'.format(vx))
70 EntryAccel.insert(0,'{:.2f}'.format(ay))
71 # ...................................................... Main loop
72 iter=0
73 while RunAll:
74   # ............................................. Draw ball on canvas
75   canvas.delete(ALL)
76   canvas.create_oval(x-rad,ch-(y+rad),x+rad,ch-(y-rad),fill=color)
77   canvas.update()
78   # ................................................ Ball is moving
79   if RunMotion:
80     # .................................................... Bouncing
81     if (x+rad)>=cw:
82       vx=-abs(vx)
83     elif (y+rad)>=ch:
84       vy=-abs(vy)
85     elif x<=rad:
86       vx=abs(vx)
87     elif y<=rad:
88       vy=abs(vy)
89     # ............. Update position and velocity, leapfrog algorithm
90     x+=vx
91     vy+=0.5*ay
92     y+=vy
93     vy+=0.5*ay
```

Lines 91–93 code Eq. (7.5) for the present case: constant acceleration and $\Delta t = 1$.

```
94   # ................................................... Read entries
95   elif GetData:
96     try:
97       vx=float(EntryVx.get())
98     except ValueError:
99       pass
100    try:
101      ay=float(EntryAccel.get())
102    except ValuError:
103      pass
104    EntryVx.delete(0,'end')
105    EntryVx.insert(0,'{:.2f}'.format(vx))
106    EntryAccel.delete(0,'end')
107    EntryAccel.insert(0,'{:.2f}'.format(ay))
108    GetData=False
```

```
    # ............................................... Write energy
    iter+=1
    if iter >=10:
      iter=0
      ener=mass*(0.5*(vx**2+vy**2)-ay*y)
      EnerLab['text']='{:8.3f}'.format(ener)
    # ............................................ Wait delay time
    canvas.after(delay)
    #——————————————————
root.destroy()
```

Running Listing 7.9 shows that now energy is conserved.

7.6.3 *The* odeint *Integration*

As a relatively simple example, we consider the elastic-string pendulum of Fig. 7.4: a bob of mass m is bound to the pivot O by a massless rubber string of rest length ℓ in the presence of standard gravity acceleration $\boldsymbol{g}$. We restrict ourselves to a motion occurring in a vertical plane. Differently from the case of Sect. 5.4, here the position of the bob cannot be specified by a single coordinate, such as the angle ϑ. Thus we shall use the Cartesian coordinates of Fig. 7.4, whose origin O is located on the pivot, while the y axis is antiparallel to the gravity acceleration $\boldsymbol{g}$. The bob of the pendulum is subject to the gravitational force $-m\boldsymbol{g}$. Further, if the distance $\sqrt{x^2+y^2}$ of the bob from the pivot is greater than ℓ, i.e., if the bob is located outside of the dashed circle of the figure, the string exerts an elastic force

$$\boldsymbol{f}_{\text{el}} = -k\left(\sqrt{x^2+y^2}-\ell\right)\left(\hat{\boldsymbol{x}}\cos\vartheta + \hat{\boldsymbol{y}}\sin\vartheta\right) , \quad \text{where} \quad \vartheta = \arctan\left(\frac{y}{x}\right) , \tag{7.6}$$

where k is the Hooke constant of the rubber band, and $\hat{\boldsymbol{x}}$ and $\hat{\boldsymbol{y}}$ are the unit vectors along the x and y axes, respectively. Obviously, the rubber band exerts no force if

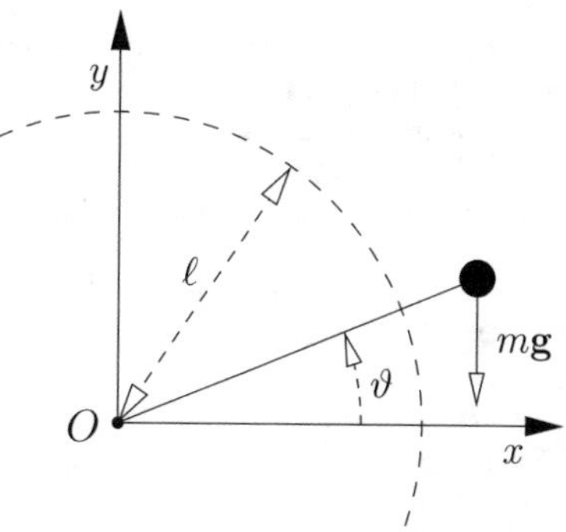

Fig. 7.4 The "elastic-string pendulum"

$\sqrt{x^2 + y^2} < \ell$ (bob inside the dashed circle). Further, we assume the presence of a frictional force proportional to the bob velocity

$$\boldsymbol{f}_{\mathrm{fr}} = -\eta\, \boldsymbol{v} \,, \tag{7.7}$$

through a given constant coefficient η. As in Sect. 5.4, we shall solve the differential equations by the function `odeint()`. Our two-dimensional problem requires the integration of the following system of four first-order differential equations

$$\begin{aligned}
\frac{\mathrm{d}x}{\mathrm{d}t} &= v_x \,, \\
\frac{\mathrm{d}v_x}{\mathrm{d}t} &= \begin{cases} -\eta v_x & \text{if } \sqrt{x^2+y^2} < \ell \,, \\ -k\left(\sqrt{x^2+y^2} - \ell\right)\cos\vartheta - \eta v_x & \text{if } \sqrt{x^2+y^2} > \ell \,, \end{cases} \\
\frac{\mathrm{d}y}{\mathrm{d}t} &= v_y \,, \\
\frac{\mathrm{d}v_y}{\mathrm{d}t} &= \begin{cases} -g - \eta v_y & \text{if } \sqrt{x^2+y^2} < \ell \,, \\ -g - k\left(\sqrt{x^2+y^2} - \ell\right)\sin\vartheta - \eta v_y & \text{if } \sqrt{x^2+y^2} > \ell \,. \end{cases}
\end{aligned} \tag{7.8}$$

In Sect. 5.4 we called `odeint()` only once, obtaining the discretization of the whole motion. Here we need to call `odeint()` before drawing each animation frame, in order to obtain an animation "in real time". Thus, while the array `t` of Listing 5.1 comprised the 101 time points at which we wanted to evaluate the bob positions, here it will comprise only two time points: the starting and the final point of the interval between two consecutive frames, arbitrarily chosen as $t_0 = 0$ and $t_1 = 0.01$ s. The listing follows.

Listing 7.10 BentBandPendulum.py

```
#!/usr/bin/env python3
from tkinter import *
import numpy as np
import time
from scipy.integrate import odeint
# ........................................... interaction functions
def StartStop():# ....... start/stop pendulum motion
  global RunIter
  RunIter=not RunIter
  if RunIter:
    StartButton['text']='Stop'
  else:
    StartButton['text']='Restart'
def ReadData(*args):# ................. read entries
  global GetData
  GetData=True
def StopAll():# ....................... exit program
  global RunAll
  RunAll=False
# ................................................. Global variables
```

```
21 RunAll=True
22 GetData=RunIter=False
```

Lines 7–19 define the three functions `StartStop()`, `ReadData()` and `StopAll()`, already encountered in the preceding listings. The program runs as long as the global variable `Runall` is *True*. The pendulum moves if the global variable `RunIter` is *True*, otherwise it is in stand-by. New data are read from the entries in the toolbar if `ReadData` is *True*.

```
23 # ............................................... Canvas data
24 ButtWidth=9
25 cw=800
26 ch=640
27 Ox=cw/2
28 Oy=ch/2
```

Variable `ButtWidth` is the maximum number of characters that can be written on a toolbar button, while `cw` and `ch` are the canvas width and height in pixels, respectively. Variables `Ox` and `Oy` are the x and y cooordinates of the pendulum pivot on the canvas, in pixels.

```
29 # ........................................... Physical parameters
30 g=9.8            # m/s^2
31 L=4.0            # m
32 m=5.0            # kg
33 Hooke=500.0      # N/m
34 eta=0.0          # kg/s
35 dt=0.01          # s
```

These are the physical quantities determining the pendulum motion, in SI units. Quantity `g` is the gravitational acceleration at the Earth's surface, in ms^{-2}, `L` is the rest length of the elastic band, in meters, `m` is the bob mass, in kg, `Hooke` is the Hooke constant of the elastic band, in N/m, `eta` is the proportionality factor η between drag force and velocity, in Ns/m=kg/s. As default we assume absence of friction, or $\eta = 0$. You can experiment interactively on how different values of η affect the motion while running the program. Finally, `dt` is the time step Δt used in the numerical integration of our differential equations.

```
36 # ..........................................................
37 prad=3           # pivot radius
38 rad=12           # bob radius
39 bColor='red'     # bob color
```

These are the parameters used for drawing the pendulum: `prad` is the pivot radius and `rad` the bob radius, in pixel, `bColor` is the bob color.

```
40 # ..........................................................
41 scale=50.0       # pixels/m
42 tau=20           # milliseconds
```

Quantity `scale` is the ratio S, expressed in px/m, of a length on the computer monitor, in px, to the corresponding real length in m. Thus the rest length of our rubber band, 4 m, corresponds to 200 pixels on the monitor. Quantity `tau` is the required time interval τ between two consecutive animation frames. Tkinter requires τ to be an integer number of milliseconds.

```
43 # ...................................... Initial position and velocity
44 xx=1.1*L
45 vx=0.0
46 yy=0.0
47 vy=0.0
48 # .................................. variable and parameter vectors
49 y=[xx,vx,yy,vy]
50 params=[L,Hooke,m,eta,g,Ox,Oy,scale,dt,tau]
```

Quantities `xx` and `yy` are the coordinates of the initial position of the bob, `vx` and `vy` the components of its initial velocity. List `y` comprises the initial conditions, at each animation step, for `odeint`.

```
51 # .................................. derivatives-computing function
52 def dfdt(y,t,params):
53   xx,vx,yy,vy = y                          # unpack initial conditions
54   L,Hooke,m,eta,g,Ox,Oy,scale,dt,tau=params    # unpack parameters
55   length=np.sqrt(xx**2+yy**2)
56   stretch=length-L
57   theta=np.arctan2(yy,xx)
58   if stretch>0:
59     force=-Hooke*stretch
60   else:
61     force=0.0
62   fx=force*np.cos(theta)-eta*vx
63   fy=force*np.sin(theta)-eta*vy
64   ax=(fx/m)
65   ay=(fy/m)-g
66   derivs=[vx,ax,vy,ay]
67   return derivs
```

Function `dfdt()` returns the derivatives needed by `odeint()` for numerical integration. Lines 53 copies the values of position and velocity at the beginning of the integration step from the list `y`. Line 54 copies the values of the parameters from the list `params`. The length of the possibly stretched rubber band, $\sqrt{x^2+y^2}$, is evaluated at Line 55, and the corresponding band extension, $\sqrt{x^2+y^2}-\ell$, at Line 56. Line 57 evaluates the angle ϑ between the string and the x axis. Lines 58–63 evaluate the x and y components, f_x and f_y, of the combined force acting on the bob due to drag and Hooke's law, according to (7.8), Lines 64–65 evaluate the corresponding acceleration components, taking also gravity into account. The derivatives of the elements of the list `y` are stored into the array `derivs` at Line 66. At Line 67 the function `dfdt()` returns `derivs`.

```
68 # ............................................. Create root window
69 root=Tk()
70 root.title('Elastic-Band Pendulum')
71 root.bind('<Return>',ReadData)
72 # ....................................... Add canvas to root window
73 canvas=Canvas(root,width=cw,height=ch,background='#ffffff')
74 canvas.grid(row=0,column=0)
```

```
75 # ........................................ Add toolbar to root window
76 toolbar=Frame(root)
77 toolbar.grid(row=0,column=1,sticky=N)
```

Lines 69–71 create the root window `root`, write the title and bind the Return keyboard key to the function `ReadData()`. Lines 73–77 create the canvas where our animation will take place, and the toolbar where buttons and entries will be located. The toolbar is located at the right of the canvas.

```
78 # .................................................... Toolbar buttons
79 nr=0
80 StartButton=Button(toolbar,text='Start',command=StartStop,\
81   width=ButtWidth)
82 StartButton.grid(row=nr,column=0,sticky=W)
83 nr+=1
84 ExitButton=Button(toolbar, text='Exit', command=StopAll,
85                   width=ButtWidth)
86 ExitButton.grid(row=nr,column=0,sticky=W)
87 nr+=1
```

Lines 79–87 create the buttons bound to the functions `StartStop()` and `StopAll()`.

```
88 # ............................................ Label and Entry arrays
89 LabVar=[]
90 EntryVar=[]
91 VarList=['x\u2080','vx\u2080','y\u2080','vy\u2080']
92 nVar=len(VarList)
93 LabPar=[]
94 EntryPar=[]
95 ParList=['Length','Hooke','Mass','\u03B7','g','Ox','Oy',\
96   'scale','Time step','\u03C4/ms']
97 nPar=len(ParList)
```

The lists that will contain the labels and entries for the variables and parameters of the problem, `LabVar`, `EntryVar`, `LabPar`, and `EntryPar` are created as initially empty lists at Lines 89, 90, 93 and 94. The lists of the variable and parameter names, `VarList` and `ParList`, to be used in the labels, are created at Lines 91 and 95. UTF-8 encoded subscripts are used for the elements of `VarList`, see Table E.1 of Appendix E. The variable names are actually x_0 (the initial x position), vx_0 (the initial x velocity component), y_0 (the initial y position), and vy_0 (the initial y velocity component). The variables of the program are the components of the bob position, x and y, and of the bob velocity, v_x and v_y. The parameters are the rest length of the rubber string, ℓ, the Hooke constant of the rubber band, k, the mass of the pendulum bob, m, the friction coefficient, η (UTF-8 code `\u03B7`), the gravity acceleration, g, the coordinates of the pivot with respect to the window frame, O_x and O_y, the length scale, the time step, Δt, and the interval between two consecutive frames, τ (UTF-8 code `\u03C4`), see Fig. 7.6. Variables `nVar` and `nPar` are the numbers of variables and parameters of the problem, respectively.

```
98 # ............................................................ Entries
99 for i in range(nVar):
```

```
100      LabVar.append(Label(toolbar,text=str(VarList[i]),\
101        font=('Helvetica',12)))
102      LabVar[i].grid(row=nr,column=0)
103      EntryVar.append(Entry(toolbar,bd =5,width=ButtWidth))
104      EntryVar[i].grid(row=nr,column=1)
105      nr+=1
106    for i in range(nPar):
107      LabPar.append(Label(toolbar,text=str(ParList[i]),\
108        font=('Helvetica',12)))
109      LabPar[i].grid(row=nr,column=0)
110      EntryPar.append(Entry(toolbar,bd =5,width=ButtWidth))
111      EntryPar[i].grid(row=nr,column=1)
112      nr+=1
```

Loop 99–105 adds the names of the variables, listed in `VarList`, to the label list `LabVar`, and locates each label at row `nr` (incremented at Line 105), column 0 of the toolbar. Then creates the entry for each variable and locates it at the right of the corresponding label, at row `nr`, column 1. Loop 106–112 adds the names of the parameters, listed in `ParList` to the parameter-label list `LabPar`, and locates the labels and the corresponding entries in the toolbar. The labels and entries for the variables and parameters are located in the toolbar as shown in Fig. 7.6.

```
113    # ................................................... time label
114    CycleLab0=Label(toolbar,text='Period:',font=('Helvetica',11))
115    CycleLab0.grid(row=nr,column=0)
116    CycleLab=Label(toolbar,text='     ',font=('Helvetica',11))
117    CycleLab.grid(row=nr,column=1,sticky=W)
118    nr+=1
119    # .............................................. Initialize entries
120    for i in range(len(VarList)):
121      EntryVar[i].insert(0,'{:.3f}'.format(y[i]))
122    for i in range(len(ParList)):
123      EntryPar[i].insert(0,'{:.3f}'.format(params[i]))
```

Lines 114–117 create two further labels, one, in column 0, with the name `Period:`, the other, in column 1, will be updated every ten iterations and show the value of the average time interval between two successive animation frames. Lines 120-123 write the initial variable and parameter values, formatted with three digits after the decimal point, into the corresponding entry windows.

```
124    # .............................................................
125    t=[0.0,dt]
126    tcount=0
127    tt0=time.time()
```

Array `t` comprises the end points of the time interval between two successive computation steps of `odeint()`. An animation frame will be drawn at each step. Variable `tcount` is a counter, that will be increased by 1 at each animation step. Function `time.time()` returns the number of seconds elapsed since the *epoch* as a float. For Unix, *epoch* is January 1st, 1970, at 0 hours. The actual value of *epoch* is usually not relevant, because only the differences between values returned by `time.time()` at different instants of the program execution are used in most programs.

```
128 # ............................................... Main loop
129 while RunAll:
130   StartIter=time.time()
131   # ........................................... Draw pendulum
132   canvas.delete(ALL)
133   canvas.create_oval(Ox-scale*L,ch-Oy+scale*L,\
134     Ox+scale*L,ch-Oy-scale*L,outline='green',width=1)
135   canvas.create_line(0,ch-Oy,cw,ch-Oy,fill='green')
136   canvas.create_oval(Ox-prad,ch-(Oy+prad),Ox+prad,ch-(Oy-prad),\
137     fill='black')
138   lengthsq=xx**2+yy**2
139   length=np.sqrt(lengthsq)
140   if length>=L:
141     canvas.create_line(Ox,ch-Oy,Ox+scale*xx,ch-Oy-scale*yy,fill='black')
142   else:
143     alpha=np.arcsin(length/L)
144     beta=np.arctan2(yy,xx)
145     gamma=(np.pi/2.0)+beta-alpha
146     xx2=0.5*L*np.cos(gamma)
147     yy2=0.5*L*np.sin(gamma)
148     canvas.create_line(Ox,ch-Oy,Ox+scale*xx2,ch-Oy-scale*yy2,\
149       Ox+scale*xx,ch-Oy-scale*yy,fill='black')
150   canvas.create_oval(Ox+scale*xx-rad,ch-Oy-scale*yy-rad,\
151     Ox+scale*xx+rad,ch-Oy-scale*yy+rad,fill=bColor)
152   canvas.update()
```

Loop 129–202 is our animation loop. Line 130 stores the initial time of the iteration (the time elapsed since *epoch*) in `StartIter`. Line 133, continued in 134, draws a green circle of radius ℓ (the rest length of the string): the elastic force is active if the center of the bob is outside the circle, inactive if the bob is inside. Line 135 draws a horizontal green line passing through the origin (through the pivot). Lines 136–137 draw a black circle in the origin, representing the pivot of the pendulum. Lines 138–139 evaluate the length of the rubber band, equal to the distance between the bob and the pivot. If the length of the rubber band is equal to, or longer than, its rest length ℓ, it is represented by a straight line drawn at Line 141. If the distance is shorter than ℓ, the rubber band is represented by a polyline comprising two line segments of equal length, starting on the pivot and ending at the center of bob, as shown in Fig. 7.5. The length of each line segment is $\ell/2$. Lines 148–149 draw the polyline. The angles α (`alpha`), β (`beta`) and γ (`gamma`), evaluated at Lines 143–145, are shown in Fig. 7.5, and are used at Lines 146 and 147 to evaluate the position (x_2, y_2) where the rubber band is bent, see Fig. 7.5. Obviously, a real rubber band would be bent in a more complicated, practically unpredictable way. Our purpose here is just to show that the distance between bob and pivot is smaller than ℓ. The bob is drawn at Lines 150–151 and the canvas is updated at Line 152.

```
153   if RunIter:
154     # .................... Velocity and position for next frame
155     psoln = odeint(dfdt,y,t,args=(params,))
156     xx=psoln[1,0]
157     vx=psoln[1,1]
158     yy=psoln[1,2]
```

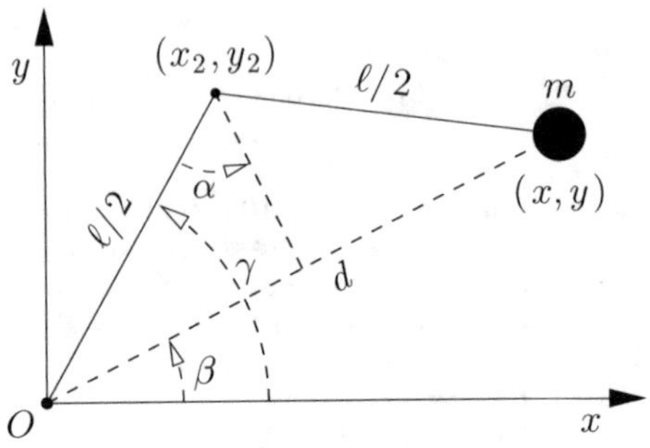

Fig. 7.5 The angles α, β and γ of Listing 7.10

```
159         vy=psoln[1,3]
160         # ............................................. Update vector
161         y=[xx,vx,yy,vy]
```

If `RunIter` is *True*, Lines 155–159 evaluate the bob position and velocity at the next step by calling `odeint()`, and the new values, x, v_x, y and v_y are stored into the list `y` at Line 161.

```
162     # ................................................. Read entries
163     elif GetData:
164         i=0
165         while i<nVar:
166             try:
167                 y[i]=float(EntryVar[i].get())
168             except ValueError:
169                 pass
170             i+=1
171         i=0
172         while i<nPar:
173             try:
174                 params[i]=float(EntryPar[i].get())
175             except ValueError:
176                 pass
177             i+=1
178         i=0
179         while i<nVar:
180             EntryVar[i].delete(0,END)
181             EntryVar[i].insert(0,'{:.3f}'.format(y[i]))
182             i+=1
183         i=0
184         while i<nPar:
185             EntryPar[i].delete(0,END)
186             EntryPar[i].insert(0,'{:.3f}'.format(params[i]))
187             i+=1
188         xx,vx,yy,vy=y
189         L,Hooke,m,eta,g,Ox,Oy,scale,dt,tau=params
190         tau=int(tau)
191         t=[0.0,dt]
192         GetData=False
```

If `RunIter` is *False* and `GetData` is *True*, Lines 163–178 read the start-values array and the parameters array from the corresponding entries in the toolbar (they may have been changed by typing new values from the keyboard). At Lines 179–187 the values are rewritten into the entries, formatted with three digits after the decimal point. Lines 188–189 copy the values from the lists. Line 190 takes care that `tau`, the required time interval between two consecutive frames, must be an integer number of milliseconds. Line 191 inserts the new value of `dt` into the list `t`. Once the new data have been read, Line 192 sets `GetData` to *False*.

```
193     # ............................................ cycle duration
194     tcount+=1
195     if tcount%10==0:
196       ttt=time.time()
197       elapsed=ttt-tt0
198       CycleLab['text']='%8.3f'%(elapsed*100.0)+' ms'
199       tt0=ttt
200     # ..........................................................
201     ElapsIter=int((time.time()-StartIter)*1000.0)
202     canvas.after(tau-ElapsIter)
```

Line 194 increases the iteration counter. Every 10 iterations Line 198 writes the average iteration duration, in ms, into label `CycleLab` at the bottom of the toolbar.

```
203     ElapsIter=int((time.time()-StartIter)*1000.0)
204     canvas.after(tau-ElapsIter)
205     #-----------------------------------------------------------
206   root.destroy()
```

Line 203 stores the time elapsed from the beginning of the current iteration, in ms, into `ElapsIter`. If `ElapsIter` is shorter than `tau` (the requested time interval between two consecutive frames), Line 204 causes the program to wait for `tau-ElapsIter` milliseconds before starting the successive iteration. Note that if `ElapsIter` is longer than `tau` Line 204 has no effect, and the program runs slower than required.

7.7 Length and Time Scaling

Script 7.10 can be instructive about some problems involved with length and time scaling in computer animation. Obviously it is convenient to use a coherent system of units when writing the differential equations to be solved by `odeint()`. If we choose SI units, lengths will be measured in meters, masses in kilograms, times in seconds, … These units are consistently used in function `dfdt()` defined at Lines 52–67, and passed to `odeint()` at Line 155. This is why the rest length of our rubber band is given in m at Line 31. But a length of 4 m would not fit into the computer monitor, and, in any case, the methods (functions) that draw items on the canvas expect lengths and coordinates to be given in px (pixels). The problem is handled by introducing the variable `scale`, corresponding to S, the number of px

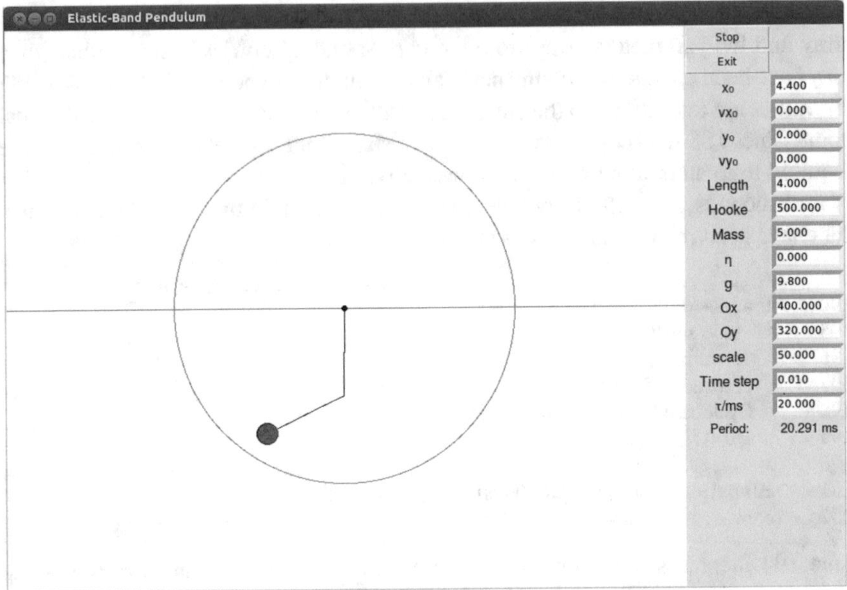

Fig. 7.6 Animation of the elastic-string pendulum. Canvas and toolbar

on the monitor that corresponds to a real length of 1 m. All real lengths and positions expressed in m must be converted into px by multiplying them by S before being passed to the Tkinter functions that draw on the canvas.

A pixel is the smallest single component of a digital image. This, on the computer display, corresponds to a small square of uniform color and of side p, the value of p depending on the specific display size and resolution of our computer. The display size is the physical size of the area where pictures and videos are displayed. The size of a screen is usually described by the length of its diagonal in inches. The display resolution is usually given in width×height with the units in pixels. For instance, on a display of size 22″ and resolution 1680 × 1050 pixels we have 96 dots per inch. Thus the side of a pixel is $p = 0.0254/96 \simeq 2.6458 \times 10^{-4}$ m.

Variable `scale` is assigned the value as $S = 50.0$ at Line 41. Thus, a real length of L, expressed in m, is represented by a line of length LS px on the monitor. Since the side of a pixel is p, the ratio of the length represented on the display to the orignal real length is

$$\rho_L = \frac{LSp}{L} = Sp. \tag{7.9}$$

Thus, the rest length of our elastic band, 4 m, corresponds to $LS = 200$ px on the canvas. With the assumed monitor size and resolution, we have $\rho_\mathrm{L} = Sp \simeq 1.3229 \times 10^{-2}$, and if we measure the radius of the green circle of Fig. 7.6 with a ruler we find

$$r = LSp \simeq 0.0529 \text{ m} = 5.29 \text{ cm}. \tag{7.10}$$

Times are measured in seconds in our computations, and Line 35 defines Variable `dt`, corresponding to the time step Δt used for the numerical solution of our differential equations (see Chap. 5), as 0.01 s, or 10 ms. On the other hand Python requires the time interval τ between two successive animation frames to be given as an integer number of milliseconds, and variable `tau` is defined at Line 42 as 20 ms. Thus the ratio of the time observed in our animation to the real time is

$$\rho_{\mathrm{t}} = \frac{1000\,\Delta t}{\tau}, \tag{7.11}$$

the factor 1000 at the numerator is due to τ being expressed in milliseconds and Δt in seconds. With the values of Script 7.10 we have $\rho_{\mathrm{t}} = 2$, and our animated motion is slower than the real motion by a factor 2. The ratio of the velocity of an object on the monitor to the corresponding real velocity is thus

$$\rho_{\mathrm{v}} = \frac{\rho_{\mathrm{L}}}{\rho_{\mathrm{t}}} = \frac{Sp\tau}{1000\,\Delta t}. \tag{7.12}$$

Again, the values of Script 7.10 lead to $\rho_{\mathrm{v}} \simeq 6.6145 \times 10^{-3}$. A real velocity of 1 m/s is represented by a velocity of 6.6145 mm/s on the monitor.

The time scale ρ_{t} can be adjusted by changing the values of Δt and/or τ. Increasing Δt, and/or decreasing τ, makes ρ_{t} larger, and movements on the display slower. Vice versa, ρ_{t} is decreased by decreasing Δt and/or increasing τ. It is important to note that the values of Δt and τ cannot be arbitrarily changed, but each of them has its own permissible range. The upper limit to Δt is set by the convergence of the finite-difference method used for the integration of the differential equations. In simple words, the values of our functions at time $t + \Delta t$ cannot be *too* different from their values at time t. The lower limit to Δt is mainly set by the computer precision, and by the animation becoming too slow. The upper limit to τ is set by the fact that a large time interval between two successive animation frames can cause the impression of a "step-wise animation". The lower limit to τ is strongly computer-dependent, being set by the computer speed. The time interval between two successive frames cannot be shorter than the time τ_{calc} needed to solve the differential equations and to redraw the canvas. If τ is set to a value shorter than τ_{calc} Line 202 simply has no effect. The reader is invited to experiment on the animation behavior when the values in the entries `Time step` and `τ/ms` are changed.

7.8 Exercises

Exercise 7.1 Write a script that displays a red square rotating about its axis. The square is centered at the center of the canvas.

Exercise 7.2 Add three buttons to the script of Exercise 7.1: one button for increasing and one for decreasing the rotation speed, and one button for exiting the program.

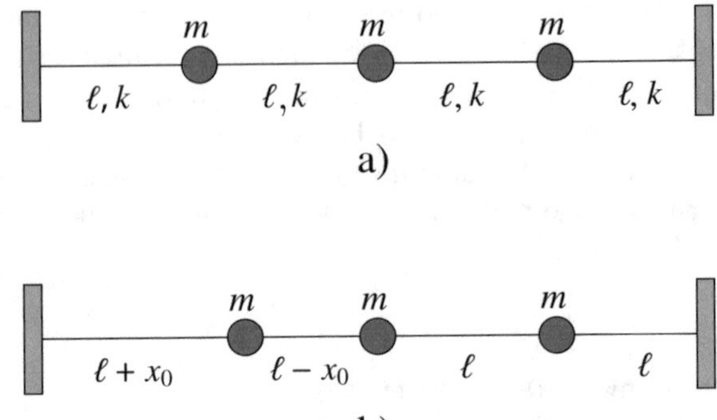

Fig. 7.7 Three coupled oscillators: **a** equilibrium positions, **b** positions at $t = 0$

Exercise 7.3 A system of coupled oscillators is formed by two external walls, three equal spheres, each of mass m, and four equal springs, each of rest length ℓ and Hooke constant k. Walls, spheres and springs are arranged as shown in Fig. 7.7a. Initially the system is at rest, with each sphere at its equilibrium position. At time $t = 0$ the first sphere is displaced by an amount x_0 from its rest position towards the second sphere, while the other two spheres are not moved, as shown in Fig. 7.7b. At time $t = 0$ the velocities of all three spheres are zero. Write a program generating the animation of the following motion. Hints: use an 800×400 px canvas, with equilibrium positions for the spheres at $x_1 = 200$ px, $x_2 = 400$ px and $x_3 = 600$ px. Assume $x_0 = 100$ px, $m = 0.1$ kg and the Hooke constants $k = 0.1$ kg/s^2.

Exercise 7.4 Add two buttons to the program you wrote for Exercise 7.3: one button for stopping and restarting the motion, and one button for exiting the program.

Exercise 7.5 Two astronomical bodies have masses $m_1 = 1.3 \times 10^{22}$ kg and $m_2 = 1.6 \times 10^{21}$ kg, respectively. At a given instant $t = 0$ the distance between the two bodies is 1.9×10^7 m, and m_2 has a velocity of 200 m/s with respect to m_1, perpendicular to the distance (these data constitute a rough approximation of the Pluto-Charon system). The gravitational constant is $G = 6.67408 \times 10^{-11}$ m^{3}kg^{-1}s^{-2}. Write a script generating an animation of the motion of the two bodies.
Hint: use barycentric coordinates, and scale the distances by a factor 2×10^{-5} px/m (good for an 800×800 px canvas).

Exercise 7.6 Write a program generating the animation of the Coulomb explosion discussed in Exercise 5.6. Hints: (i) represent the cluster of helium atoms by a circle a the center of the canvas: (ii) use a scale of 5×10^7 px/m, so that the initial radius of the cluster is 5 px; (iii) use a "real" time interval $\Delta t = 10^{-11}$ s between two successive frames of the animation.

Exercise 7.7 Add two labels to the toolbar of Exercise 7.6, one label reporting the current cluster radius in nm (nanometers), and one reporting the elapsed real time in ps (picoseconds).

Chapter 8
Classes

8.1 The `class` Statement

Python is an *object-oriented* programming (OOP) language. While *procedure-oriented* programming (POP) languages are mainly based on variables and functions, OOP languages add, and stress, *objects*. Often objects correspond to things found in the real world. A graphics program may have objects such as “circle” or “square”, a physics program may have objects such as “electron” or “nucleus”. An object is a collection of data (variables) characterizing the object itself, and methods (functions) acting on those data. It is certainly possible to write Python programs not using objects, as we have done in the previous chapters. But objects are one of the strengths of Python. Python, as most OOP languages, is class-based: Python objects are instances of *classes*. This means that the structure and behavior of an object are defined by a *class*, a class being a blueprint of all objects of a specific type. For example, a celestial-mechanics program can define *objects* “Earth” and ”Jupiter” as instances of the *class* “planet”. A class is defined by the `class` statement, ending with a colon. Its methods are defined by using the `def` statement, just as usual functions. As usual, code blocks are defined by their indentation. Listing 8.1 should give a first idea of how a class definition is structured.

Listing 8.1 ClassExample.py

```
#!/usr/bin/env python3
# .............................. class particle
class particle:
  color='red'
  def __init__(self,mass,x,vx):
    self.m=mass
    self.x=x
    self.vx=vx
    self.px=mass*vx
```

Electronic supplementary material The online version of this chapter (https://doi.org/10.1007/978-3-030-45027-4_8) contains supplementary material, which is available to authorized users.

G. Moruzzi, *Essential Python for the Physicist*,
https://doi.org/10.1007/978-3-030-45027-4_8

```
10        # ............................ move particle
11    def move(self):
12        self.x+=self.vx
13  # ..............................................
14  pt1=particle(10,10,3.5)
15  pt2=particle(20,15,-2.1)
16  center=(pt1.m*pt1.x+pt2.m*pt2.x)/(pt1.m+pt2.m)
17  print("momentum1 = {:.1f}".format(pt1.px))
18  print("momentum2 = {:.1f}".format(pt2.px))
19  print("center of mass = {:.3f}".format(center))
20  print('color1:',pt1.color,'  color2:',pt2.color)
```

Lines 3–12 comprise the definition of a class named `particle`. This class has two *methods*, one *class variable* and four *instance variables*. Line 4 defines the class variable color, which is set equal to `'red'`. A class variable is a variable which has the same value for all class instances: all our particles will be red. Lines 5–9 define the `__init__()` method, note the leading and trailing double underscores (__) reserved by Python to the names of special methods. Method `__init__()` is a function that creates a new object belonging to the class (a new class instance) in a given *initial state*, specified by the initial values of some *instance variables*. Here method `__init__()` has four arguments, the first, `self`, is a reference to the current instance of the class. The other arguments are the mass, initial position and initial velocity of a particle moving in one dimension (along the x axis). Lines 6–8 copy the argument values of `__init__()` into instance variables of the new class instance. Instance variables are variables that are specific to the object, and may differ from one instance to the other of the same class. Instance variables are prefixed by "`self.`", like `self.m`, in the class declaration. Line 9 defines the momentum of the particle. Lines 11–12 define the `move()` method which, when called, displaces the particle position by `vx`. In an animation, a loop will iteratively call the `move()` method, each time displacing the particle. Thus, here the velocity is actually measured in pixels/cycle, as in previous scripts.

Lines 14–15 create two new "particles", i.e., two new instances of the class. The first particle is called `pt1`, with `mass` $m_1 = 10$, position $x_1 = 10$, and velocity $v_1 = 3.5$, the second is called `pt2`, with $m_2 = 20$, $x_2 = 15$, and $v_2 = -2.1$. This is how class instances are created.

Line 16 evaluates the center of mass of the two particles, `center` (x_c in mathematical notation)

$$x_c = \frac{m_1x_1 + m_2x_2}{m_1 + m_2} \,. \tag{8.1}$$

Note the *dot notation* for accessing the instance variables relative to each particle: `pt1.m` indicates the mass of particle 1, `pt2.m` indicates the mass of particle 2. Lines 17 and 18 print the values of the momenta of particle 1 and particle 2. Line 19 prints the center-of-mass coordinate of the two-particle system. Line 20 prints the colors of the two particles, both red, because we have defined `color` as a class variable. This is what we see when we run the script

```
$> ClassExample.py
momentum1 = 35.0
momentum2 = -42.0
center of mass = 13.333
color1: red    color2: red
```

8.2 A First Example: Two Colliding Balls

Script 8.2 provides a first example of the use of classes in Tkinter animation. Two balls of different size and mass, both instances of the class `ball`, move on the canvas, bouncing against the canvas borders and colliding elastically with each other. As usual, a backslash (\) at the end of a code line indicates that a long command is split over multiple lines.

Listing 8.2 ClassCollide.py

```
 1  #!/usr/bin/env python3
 2  from tkinter import *
 3  from numpy import arctan2,cos,sin,sqrt
 4  import time
 5  # .............................................. Global variables
 6  RunAll=True
 7  GetData=RunIter=False
 8  ButtWidth=9
 9  # ................................................... Canvas sizes
10  cw=800
11  ch=600
12  # ................................................... Start values
13  tau=20  # milliseconds
14  m1=200
15  r1=40
16  x1=r1
17  y1=r1
18  vx1=5.0
19  vy1=5.0
20  m2=150
21  r2=30
22  x2=cw-r2
23  y2=r2
24  vx2=-5.0
25  vy2=5.0
```

Line 13 assigns the required time interval between two successive animation frames, $\tau = 20$ ms. Lines 14–25 assign the mass in arbitrary units, radius, the initial x and y coordinates in px, the initial x and y velocity components in px/cycle, for two instances of the class `ball`. As usual in simple animations, positions are measured in pixels and velocities in pixels/(animation cycle).

```
26  # ...................................................... Class ball
27  class ball:
28    def __init__(self,mass,radius,x,y,vx,vy,color):
29      self.m=mass
30      self.rad=radius
```

```
31         self.x=x
32         self.y=y
33         self.vx=vx
34         self.vy=vy
35         self.col=color
36         self.image=canvas.create_oval(self.x-self.rad,ch-(self.y+\
37           self.rad),self.x+self.rad,ch-(self.y-self.rad),\
38             fill=self.col,outline=self.col)
```

Lines 27–108 define the class `ball`. The `__init__()` method is defined at lines 28–38: each instance of this class will have its own mass, radius, position and velocity on the canvas plane, and its own color. Lines 36–38 create a first image of the class instance on the canvas, whose position can later be changed by the methods `canvas.coords()` and `canvas.move()`. This is an alternative to what seen in Script 7.10: instead of clearing the whole canvas with the command `canvas.delete(ALL)` at Line 132, and redrawing everything, here each object will be moved separately on the canvas. As usual in physics, we shall use an upwards-directed y axis for computations. Conversion to the canvas coordinates for drawing is performed by writing `ch-self.y` instead of `self.y`. Variable `ch` is global and corresponds to the canvas height in px.

```
39       # ................................................ Move ball
40       def move(self):
41         self.x+=self.vx
42         self.y+=self.vy
43         canvas.coords(self.image,self.x-self.rad,ch-(self.y+self.rad),\
44           self.x+self.rad,ch-(self.y-self.rad))
```

Method `move()`, defined at Lines 40–44, moves the ball to its position in the successive animation frame. First, the x and y coordinates of its center are updated at lines 41–42, then, the position of `self.image` is updated by `canvas.coords()` at lines 43–44. The first argument of the canvas method `.coords()` is the object to be relocated, in this case `self.image`, the image of the class instance (the current ball) on the canvas. The other arguments are coordinates specifying the new position of the object. The number of coordinates depends on the object. Here they have the form x_1, y_1, x_2, y_2, describing the bounding box of the oval representing the ball.

```
45       # ................................... Bounce on canvas borders
46       def bounce(self):
47         if (self.x+self.rad)>=cw:
48           self.vx=-abs(self.vx)
49           self.x=2.0*(cw-self.rad)-self.x
50         if (self.x-self.rad)<=0:
51           self.vx=abs(self.vx)
52           self.x=2.0*self.rad-self.x
53         if (self.y+self.rad)>=ch:
54           self.vy=-abs(self.vy)
55           self.y=2.0*(ch-self.rad)-self.y
56         if (self.y-self.rad)<=0:
57           self.vy=abs(self.vy)
58           self.y=2.0*self.rad-self.y
```

Lines 46–58 define the method `bounce()`, which checks if the ball, at its current position, has reached one of the canvas borders or trespassed it. Trespassing can happen (mathematically, not physically!) because the ball moves by discrete steps. In this case the method makes the ball bounce at the canvas border. Lines 47 checks if the x position plus the ball radius is beyond the canvas right border. If so, the sign of the x velocity component is made negative at Line 48, and the x coordinate is reflected on the canvas border at Line 49. Lines 50–58 perform analogous checks and transformations at the other three canvas borders.

```
59      # ........................................ Elastic collision
60    def ElastColl(self,other):
61      X=other.x-self.x
62      Y=other.y-self.y
63      distsq=X**2+Y**2
64      R12sq=(self.rad+other.rad)**2
65      if distsq<=R12sq:
66        tc=0.0
67        # .................................. Adjust overlapping balls
68        if distsq<R12sq:
69          Xdot=other.vx-self.vx
70          Ydot=other.vy-self.vy
71          aa=Xdot**2+Ydot**2
72          bbhalf=X*Xdot+Y*Ydot
73          cc=X**2+Y**2-R12sq
74          # ...................... Time elapsed since "real" collision
75          tc=(-bbhalf-sqrt(bbhalf**2-aa*cc))/aa
76          # ....................... Time reversal to collision instant
77          other.x+=tc*other.vx
78          other.y+=tc*other.vy
79          self.x+=tc*self.vx
80          self.y+=tc*self.vy
81          # .......................... Distances at collision instant
82          X=other.x-self.x
83          Y=other.y-self.y
84        # ................................. Collision reference frame
85        alpha=arctan2(Y,X)
86        csalpha=cos(alpha)
87        snalpha=sin(alpha)
88        SelfVelXi=self.vx*csalpha+self.vy*snalpha
89        SelfVelEta=-self.vx*snalpha+self.vy*csalpha
90        OtherVelXi=other.vx*csalpha+other.vy*snalpha
91        OtherVelEta=-other.vx*snalpha+other.vy*csalpha
92        SelfNewVelXi=((self.m-other.m)*SelfVelXi+2.0*other.m*OtherVelXi)/\
93          (self.m+other.m)
94        OtherNewXi=((other.m-self.m)*OtherVelXi+2.0*self.m*SelfVelXi)/\
95          (self.m+other.m)
96        self.vx=SelfNewVelXi*csalpha-SelfVelEta*snalpha
97        self.vy=SelfNewVelXi*snalpha+SelfVelEta*csalpha
98        other.vx=OtherNewXi*csalpha-OtherVelEta*snalpha
99        other.vy=OtherNewXi*snalpha+OtherVelEta*csalpha
100       # ........................................................
101       other.x-=tc*other.vx
102       other.y-=tc*other.vy
```

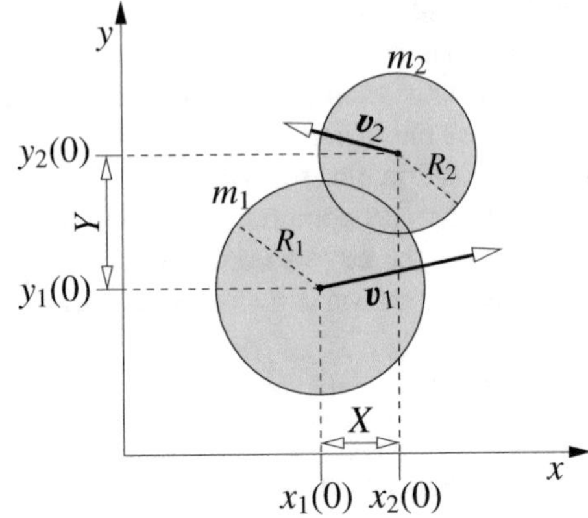

Fig. 8.1 An overlapping of the colliding spheres can occur in the calculated motion, because the calculations are performed stepwise

```
103         self.x-=tc*self.vx
104         self.y-=tc*self.vy
```

Lines 60–104 define the method `ElastColl()`, that handles elastic collisions between the specific class instance (the specific ball, let us call it *ball* 1) and another class instance (another ball, let us call it *ball* 2). The method's second argument, `other`, refers to ball 2, thus, it must be another instance of the class `ball`. Lines 61–64 evaluate variables `X`, defined as $X = x_2 - x_1$, and `Y`, defined as $Y = y_2 - y_1$. Variable `distsq` is the square of the distance between the centers of the two balls, $X^2 + Y^2$, and `R12sq` is the square of the sum of the two radii, $(R_1 + R_2)^2$. If $X^2 + Y^2 > (R_1 + R_2)^2$ no collision occurs, and the method is exited at Line 65. Otherwise variable `tc`, corresponding to the time interval t_c elapsed from the collision instant to the time of the current animation frame (the two balls might be compenetrating because of their stepwise motions), is temporarily set to 0.

Line 68 checks if $X^2 + Y^2 < (R_1 + R_2)^2$, the other remaining possibility being $X^2 + Y^2 = (R_1 + R_2)^2$. In the latter case the two balls are tangent to each other, and we are exactly at the collision instant. Otherwise the two balls are partially overlapping as shown in Fig. 8.1. As mentioned above, the two balls cannot overlap in reality, but they can overlap in our simulation because they move by discrete steps. Lines 69–83 of our listing handle this case by performing a "time reversal" that brings the balls back in time to the collision instant, which occurred between the times of the two successive animation frames. The coordinates of ball i, where $i = 1, 2$, can be written as functions of time as

$$x_i(t) = x_i(0) + \dot{x}_i t \ , \quad y_i(t) = y_i(0) + \dot{y}_i t \ , \tag{8.2}$$

where $x_i(0)$ and $y_i(0)$ are the calculated coordinates of the overlapping balls at the current animation frame instant, that we choose as time origin, while $\dot{x}_i$ and $\dot{y}_i$ are their velocity components. Thus, for instance, $\dot{x}_1$ corresponds to the variable `self.vx`. Remember that time is measured in units of τ, the interval between between two

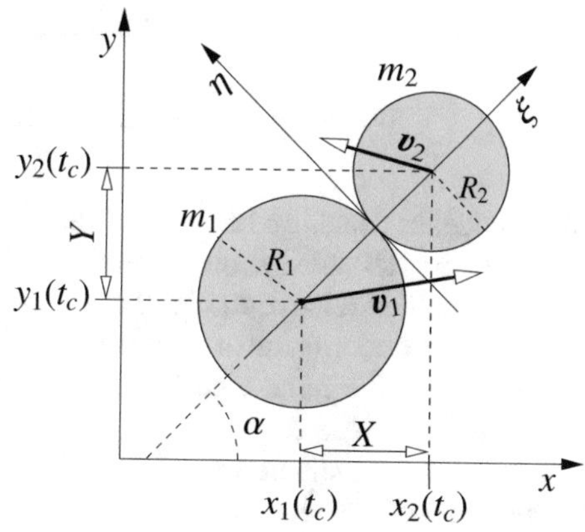

Fig. 8.2 The two spheres at the collision instant

successive frames. Accordingly, the square of the distance between the two centers, $s^2(t)$, is written, as a function of time,

$$\begin{aligned} s^2(t) &= [x_2(t) - x_1(t)]^2 + [y_2(t) - y_1(t)]^2 \\ &= \big[(x_2(0) + \dot{x}_2 t - x_1(0) - \dot{x}_1 t)^2 + \\ &\qquad (y_2(0) + \dot{y}_2 t - y_1(0) - \dot{y}_1 t)^2\big] \\ &= \left(\dot{X}^2 + \dot{Y}^2\right) t^2 + 2\left(X\dot{X} + Y\dot{Y}\right) t + X^2 + Y^2 , \end{aligned} \tag{8.3}$$

where $\dot{X} = \dot{x}_2 - \dot{x}_1$ and $\dot{Y} = \dot{y}_2 - \dot{y}_1$. In order to find the collision time t_c, such that $s^2(t_c) = (R_1 + R_2)^2$, we must solve the quadratic equation

$$\left(\dot{X}^2 + \dot{Y}^2\right) t_c^2 + 2\left(X\dot{X} + Y\dot{Y}\right) t_c + \left[X^2 + Y^2 - (R_1 + R_2)^2\right] = 0 . \tag{8.4}$$

This is done at lines 69–75, where, at Line 75, we choose the negative solution for `tc` because, obviously, the collision occurred before the balls overlapped. Note that we must have $|t_c| < 1$ because the collision occurred between the times of the two successive frames, separated by our time unit τ. Lines 77–80 evaluate the coordinates of the ball centers at the collision instant t_c, and Lines 82–83 the new variables $X = x_2(t_c) - x_1(t_c)$ and $Y = y_2(t_c) - y_1(t_c)$

The collision instant of two spheres of masses m_1 and m_2, radii R_1 and R_2, centers located at $[x_1(t_c), y_1(t_c)]$ and $[x_2(t_c), y_2(t_c)]$, and velocities $\boldsymbol{v}_1$ and $\boldsymbol{v}_2$, respectively, is shown in Fig. 8.2. In a two-dimensional elastic collision between two perfectly smooth spheres, according to classical mechanics, the velocity of each sphere must be split into two perpendicular components: one tangent to the surfaces of the colliding spheres at the point of contact (along the η axis in Fig. 8.2), the other along the line joining the centers of the two spheres (along the ξ axis). Assuming that the sphere surfaces are perfectly smooth, the collision affects only the ξ velocity components, while the η components are unchanged. Denoting by $v_{1\xi}$ and $v_{2\xi}$ the ξ velocity components of the two spheres before the collision, the ξ velocity components after the collision, $w_{1\xi}$ and $w_{2\xi}$, are written

$$
\begin{aligned}
w_{1\xi} &= \frac{(m_1 - m_2)\, v_{1\xi} + 2\, m_2\, v_{2\xi}}{m_1 + m_2}\,, \\
w_{2\xi} &= \frac{(m_2 - m_1)\, v_{2\xi} + 2\, m_1\, v_{1\xi}}{m_1 + m_2}\,.
\end{aligned} \tag{8.5}
$$

In the code, as usual, `self` refers to the specific class instance (ball 1), while argument `other`, as stated above, refers to the other ball involved in the collision, ball 2. The variables relative to ball 2 are accessed through the *dot notation*, thus, for instance, the x coordinate of its center is `other.x`. Lines 82–83 evaluate the differences between the x and y positions of the two balls, `X` and `Y`, respectively, denoted by X and Y in Fig. 8.2.

Lines 85–87 evaluate the angle α (`alpha`) between the $\xi\eta$ and xy reference frames, its cosine (`csalpha`) and sine (`snalpha`). Lines 88 and 89 evaluate the ξ and η velocity components of ball 1 before the collision, i.e., quantities $v_{1\xi}$ and $v_{1\eta}$ of (8.5), stored in `SelfXi` and `SelfEta`, respectively. Lines 90 and 91 evaluate the corresponding quantities $v_{2\xi}$ and $v_{2\eta}$ for ball 2, stored in `OtherXi` and `OtherEta`. Lines 92–93 evaluate the ξ velocity component of ball 1 after the collision, $w_{1\xi}$ stored in `SelfNewXi`, according to (8.5). Lines 94–95 store the ξ velocity component of ball 2 after the collision into `OtherNewXi`. Finally, Lines 96–99 evaluate the final velocity components of both balls in the "laboratory frame" xy, i.e., `self.vx`, `self.vy`, `other.vx`, and `other.vy`.

Lines 101–104 take into account that, since the collision took place at time $t_c < 0$, thus before the instant of the current animation frame, at $t = 0$ the two balls have moved further for a time $|t_c|$ with their new velocities.

```
105 # ............................................. Button functions
106 def ReadData(*arg):
107     global GetData
108     GetData=True
109 #
110 def StartStop():
111     global RunIter
112     RunIter=not RunIter
113     if RunIter:
114         StartButton["text"]="Stop"
115     else:
116         StartButton["text"]="Restart"
117 #
118 def StopAll():
119     global RunAll
120     RunAll=False
```

Functions `ReadData()`, `StartStop()` and `StopAll()` are the same as in Script 7.10.

```
121 # ............................................. Create root window
122 root=Tk()
123 root.title('Class Collide')
124 root.bind('<Return>',ReadData)
125 # ...............................................................
```

```
126  canvas=Canvas(root,width=cw,height=ch,background="#ffffff")
127  canvas.grid(row=0,column=0)
128  # ............................................ Create toolbar
129  toolbar=Frame(root)
130  toolbar.grid(row=0,column=1, sticky=N)
```

Lines 122–124 create the root window `root`, write the title in its frame, and bind the Return keyboard key to the function `ReadData()`. Lines 126–130 create the canvas and the toolbar. The toolbar, that will contain the buttons and entries, is located at the right of the canvas

```
131  # ........................................................ Buttons
132  nr=0
133  StartButton=Button(toolbar,text="Start",command=StartStop,
134                     width=ButtWidth)
135  StartButton.grid(row=nr,column=0,sticky=W)
136  nr+=1
137  CloseButton=Button(toolbar, text="Exit", command=StopAll,
138                     width=ButtWidth)
139  CloseButton.grid(row=nr,column=0,sticky=W)
140  nr+=1
```

Lines 132–140 create the buttons bound to the functions `StartStop()` and `StopAll()`.

```
141  # ............................................... Parameter arrays
142  LabPar=[]
143  EntryPar=[]
144  ParList=['m\u2081','r\u2081','vx\u2081','vy\u2081','m\u2082','r\u2082',
145           'vx\u2082','vy\u2082','\u03C4']
146  nPar=len(ParList)
147  # ............................... Entries for new parameter values
148  for i in range (nPar):
149    LabPar.append(Label(toolbar,text=str(ParList[i]),
150                        font=("Helvetica",12)))
151    LabPar[i].grid(row=nr,column=0)
152    EntryPar.append(Entry(toolbar,bd=5,width=10))
153    EntryPar[i].grid(row=nr,column=1)
154    nr+=1
```

Lines 142–146 create the lists that will contain the labels and entries for the parameters of the problem, `LabPar` and `EntryPar`, as well as the list of the parameter names, `ParList`. For the elements of `ParList`, UTF-8 encoded subscripts are used, see Table E.1 of Appendix E. Lines 148–150 locate labels and entries on the toolbar as shown in Fig. 8.3.

```
155  # ..................................................... Time label
156  CycleLab0=Label(toolbar,text="Period:",font=("Helvetica",11))
157  CycleLab0.grid(row=nr,column=0)
158  CycleLab=Label(toolbar,text="␣␣␣␣␣",font=("Helvetica",11))
159  CycleLab.grid(row=nr,column=1,sticky=W)
160  nr+=1
```

Label `CycleLab` will show the actual duration of the time interval between two successive animation frames.

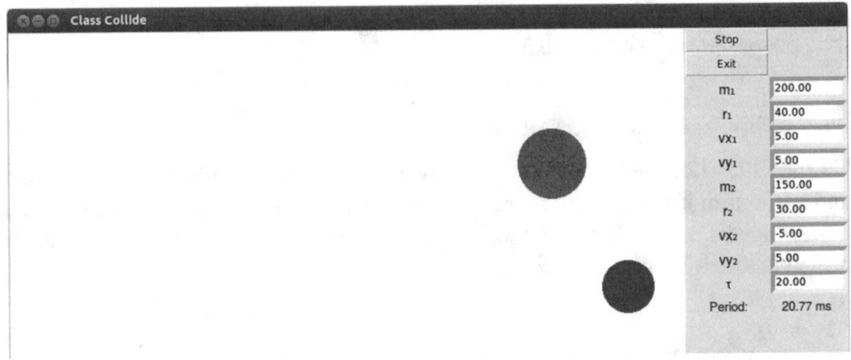

Fig. 8.3 The two balls moving on the canvas, and the toolbar with buttons and entries

```
161 # ..................................................... Parameters
162 params=[m1,r1,vx1,vy1,m2,r2,vx2,vy2,tau]
163 for i in range(nPar):
164   buff="%.2f" % params[i]
165   EntryPar[i].delete(0,'end')
166   EntryPar[i].insert(0,buff)
```

The parameter list `params` is created, and the initial values of the parameters are written in the windows of the toolbar entries.

```
167 # ........................................ Create colliding balls
168 ball1=ball(m1,r1,x1,y1,vx1,vy1,"red")
169 ball2=ball(m2,r2,x2,y2,vx2,vy2,"blue")
```

The two collision partners `ball1` and `ball2` are created as instances of the class `ball`, each with its own mass, initial position and velocity, and color.

```
170 # ..................................................... Time origin
171 tt0=time.time()
172 tcount=0
```

The initial time `tt0` and the counter `tcount are initialized.`

```
173 # .................................................. Animation loop
174 while RunAll:
175   StartIter=time.time()
176   # ................................................. Move balls
177   if RunIter:
178     ball1.move()
179     ball1.bounce()
180     ball2.move()
181     ball2.bounce()
182     ball1.ElastColl(ball2)
```

Lines 174–223 comprise the animation loop. Line 175 stores the start time of each iteration into `StartIter`. If `RunIter` is *True*, lines 178–182 move each ball, check if the balls are bouncing at the canvas borders and check if an elastic collisions is occurring between them. Note that the command `ball2.ElastColl(ball1)` would be equivalent to `ball1.ElastColl(ball2)`.

```
183 else:
184     if GetData:
185         i=0
186         while i<nPar:
187             try:
188                 params[i]=float(EntryPar[i].get())
189             except ValueError:
190                 pass
191             i+=1
192         ball1.m,ball1.rad,ball1.vx,ball1.vy,\
193             ball2.m,ball2.rad,ball2.vx,ball2.vy,tau=params
194         tau=int(tau)
195         for i in range(nPar):
196             buff="%.2f" % params[i]
197             EntryPar[i].delete(0,'end')
198             EntryPar[i].insert(0,buff)
199         ball1.x=ball1.rad-ball1.vx
200         ball1.y=ball1.rad-ball1.vy
201         ball2.x=cw-ball2.rad-ball2.vx
202         ball2.y=ball2.rad-ball2.vy
203         ball1.move()
204         ball2.move()
205         GetData=False
```

If `Runiter` is *False*, and if `GetData` is *True*, the loop 186–191 updates the list `params` reading the new values from the entries on the toolbar. The single variables are read from the `params` array at Lines 192–193. Variable `tau` is converted to an integer number of milliseconds at Line 194.

Loop 195–198 rewrites the entry windows, formatting the values with two digits after the decimal point. Lines 199–200 locate the center of ball 1 at $x = r_1 - v_{x1}$, $y = r_1 - v_{y1}$. Thus, when the method `ball1.move()` is called at Line 203, the ball center will be located at $x = r_1$, $y = r_1$. The same is done for ball 2 at Lines 201–202. Line 205 sets `GetData` to *False*, so that entries are no longer read till the Enter key is pressed again.

```
206 # ............................................. Cycle duration
207 tcount+=1
208 if tcount==10:
209     tcount=0
210     ttt=time.time()
211     elapsed=ttt-tt0
212     CycleLab['text']="%8.2f"%(elapsed*100.0)+" ms"
213     tt0=ttt
214 ElapsIter=int((time.time()-StartIter)*1000.0)
215 canvas.update()
216 canvas.after(tau-ElapsIter)
```

At Line 207 the counter `tcount` is increased by 1. When `tcount` equals 10 Line 211 determines the duration time of the last 10 loop iterations, and Line 212 prints the average duration of an iteration, in ms, in the label CycleLab at the bottom of the toolbar in Fig. 8.3.

Line 214 measures the time elapsed since the beginning of the current iteration. Line 215 updates the canvas, and Line 216 delays the next iteration till a total time τ has elapsed between two successive animation frames.

```
217   #———————————————————————————————————————
218   root.destroy()
```

8.3 A "Classical" Atom

In physics books we learn that a "classical" atom, i.e., an atom comprising a nucleus and electrons obeying the laws of classical electromagnetism, would not be stable. Such an atom would collapse because the orbiting electrons, being accelerated, would radiate energy and spiral down to the nucleus. We also learn that this does not occur because, actually, atoms obey the laws of quantum mechanics rather than the laws of classical physics. And this is the end of the story.

However, it is interesting to note that a hypothetical "classical" atom with two or more electrons, thus, any classical atom more complex than the hydrogen atom, would not be stable even disregarding radiation losses. Instinctively we might think of a classical atom as something similar to the Solar System, with the nucleus playing the role of the Sun and the electrons the roles of the planets. But there is a very important difference. The gravitational force between any two components of the Solar System (two planets, or a planet and the Sun) is proportional to the product of the two masses, with the mass of the Sun (1.99×10^{30} kg) being more than 1000 times larger than the mass of the most massive planet, Jupiter (1.90×10^{27} kg). This makes neglecting the interactions of the planets between themselves a reasonable start approximation. And, in any case, one ought to remember that it took to the Solar System some 4 billion years to reach the actual "stable" configuration.

On the other hand, it is true that a "classical" helium atom (our simplest example) would be a three-body system where the mass of the nucleus is much larger than the mass of each electron (by a factor of approximately 7.3×10^3), in analogy to the Solar System. But here forces are proportional to the products of the involved charges, instead of masses. Thus, the interaction between the two electrons is absolutely not negligible with respect to the nucleus-electron interactions. This makes approximations analogous to the ones used for the Solar System impossible, and an approximate analytic treatment of the problem is not feasible. Even if, obviously, a classical atom simply does not exist, it can be instructive to watch the animation generated by Script 8.3.

Listing 8.3 ClassicalAtom.py

```
#!/usr/bin/env python3
#
from tkinter import *
```

```
4  from scipy.integrate import odeint
5  import numpy as np
6  import time
7  # ......................................... subscripts for labels
8  sub=['\u2080','\u2081','\u2082','\u2083','\u2084','\u2085',\
9     '\u2086','\u2087','\u2088','\u2089']
10 # ............................................... Global variables
11 RunAll=True
12 RunIter=NewBaryc=GetData=ReWrite=False
13 # ................................................ Physical values
14 q=1.602176e-19          # elementary charge/Coulomb
15 me=9.10938e-31          # electron mass/kg
16 mp=1.67262e-27          # proton mass/kg
17 ke=8.987551e9           # Coulomb's constant (N m^2/C^2)
18 r2=1.0e-10              # radius of second-electron orbit / m
19 r1=r2/3.0
20 v1=np.sqrt(ke*2.0*q**2/(me*r1)) # m/s
21 v2=np.sqrt(ke*q**2/(me*r2))     # m/s
22 dt=2.0e-19              # s
```

The list at Lines 8–9 defines numerical subscripts according to Table E.1 of Appendix E. These subscripts will be used in the the labels that appear in the toolbar, see Fig. 8.4. Lines 11–12 initialize the usual global variables needed by the toolbar buttons.

Lines 14–17 define some constants in SI units: the elementary charge q, the electron and proton masses m_{e} and m_{p}, and the Coulomb constant $k_{\mathrm{e}} = 1/(4\pi\varepsilon_0)$. As initial conditions we arbitrarily assume all particles (nucleus and electrons), located on the x axis of a Cartesian reference frame, the nucleus being located at the origin. Variable `r2` is the initial distance $r_2 = x_2^0$ of the farther electron from the nucleus, arbitrarily chosen as 0.1 nm. Variable `r1` is the initial distance of the closer electron, $r_1 = -x_1^0$, from the nucleus. We arbitrarily choose $x_1^0 = x_2^0/3$. The initial velocities of both electrons are assumed parallel to the y axis, $\boldsymbol{v}_1^0 = (0, \dot{y}_1)$ and $\boldsymbol{v}_2^0 = (0, \dot{y}_2)$. Variable `v1`, corresponding to $\dot{y}_1$, is chosen so that the closer electron would describe a circular orbit around the nucleus in the absence of electron 2. Analogously, the initial velocity of the farther electron, `-v2`, corresponding to $\dot{y}_2$, would correspond to a circular orbit if electron 1 had collapsed onto the nucleus. Finally, `dt` is the step size $\mathrm{d}t$, in seconds, used for the numerical integration of the equations of motion.

```
23 # .................................. Drawing and animation parameters
24 cycle=20                # ms
25 scale=3.0e12            # px/m
26 cw=900                  # px
27 ch=900                  # px
28 Ox=cw/2.0
29 Oy=ch/2.0
30 bcrad=2                 # px
31 TrailLength=200
```

Quantity `cycle` is the *required* time interval $\Delta\tau$ between two successive animation frames. Quantity `scale` is the ratio S between a distance on the canvas, measured in px, and the corresponding real distance measured in m. Thus, a real

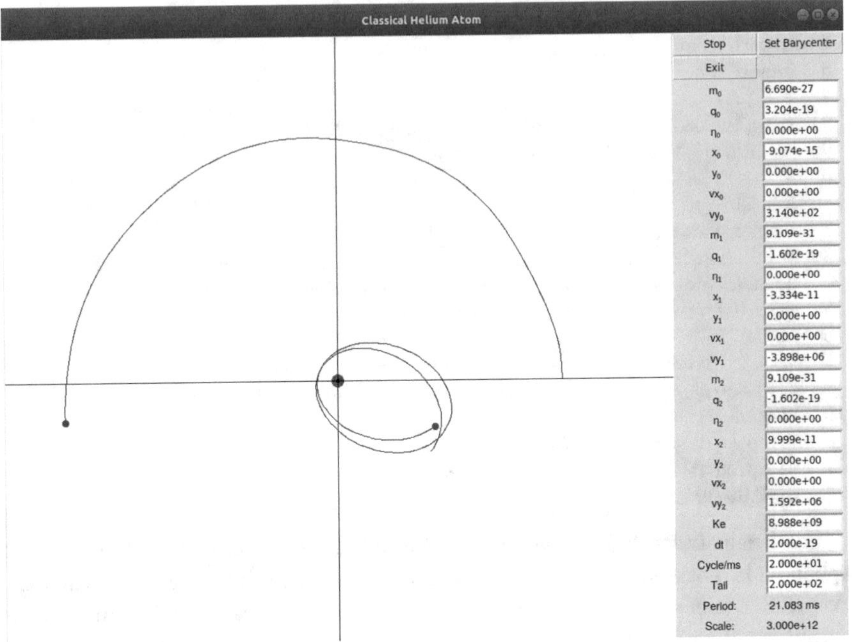

Fig. 8.4 Electron orbits in a "classical" helium atom

distance of 1 m corresponds to 3×10^{12} px, a distance of 0.1 nm corresponds to 300 px on the canvas. As we shall see in the following, the value of `scale` can be interactively changed during the program execution.

As usual, `cw` and `ch` are the canvas width and height in px, respectively, while `Ox` and `Oy` are the position, in px, of the origin of our coordinate system with respect to the left upper corner of the canvas. The barycenter of the atom will be shown by a small black circle of radius 2 px (`bcrad`). In order to visualize the paths of the particles, each particle (electron or nucleus) leaves a trail on the canvas, consisting of a polyline of `TrailLength` (N_{trail} in mathematical notation) vertices, starting from the present position of the particle, as shown in Fig. 8.4.

```
# ............................................. Start/Stop function
def StartStop():
  global RunIter
  RunIter=not RunIter
  if RunIter:
    StartButton["text"]="Stop"
  else:
    StartButton["text"]="Restart"
# ................................................... Exit function
def StopAll():
  global RunAll
  RunAll=False
# ............................................ Read Data function
```

```
45 def ReadData(*arg):
46   global GetData
47   GetData=True
48 # ................................................. Scale Down
49 def ScaleDown(*arg):
50   global scale
51   scale/=np.sqrt(2.0)
52   ScaleLab['text']="%10.3e"%(scale)
53 # ................................................. Scale Up
54 def ScaleUp(*arg):
55   global scale
56   scale*=np.sqrt(2.0)
57   ScaleLab['text']="%10.3e"%(scale)
```

Functions `StartStop()`, `StopAll()` and `ReadData()` are the same as in Script 7.10. Function `ScaleDown()` divides `scale` by a factor $\sqrt{2}$ each tim it is called, thus reducing the lengths on the canvas. Function `ScaleUp()` muliplies `scale` by $\sqrt{2}$, thus enlarging the picture.

```
58 # ......................... Evaluate center of mass and its velocity
59 def baryc(part):
60   mtot=sum(zz.m for zz in part)
61   cx=sum(zz.x*zz.m for zz in part)/mtot
62   cy=sum(zz.y*zz.m for zz in part)/mtot
63   cvx=sum(zz.vx*zz.m for zz in part)/mtot
64   cvy=sum(zz.vy*zz.m for zz in part)/mtot
65   return [[cx,cy],[cvx,cvy]]
66 # .................................. move origin to center of mass
67 def SetBaryc():
68   global NewBaryc
69   global part
70   xcm,ycm=baryc(part)[0]
71   cvx,cvy=baryc(part)[1]
72   for zz in part:
73     zz.x-=xcm
74     zz.y-=ycm
75     zz.vx-=cvx
76     zz.vy-=cvy
77   NewBaryc=True
```

Function `baryc()` evaluates the position and velocity components of the barycenter of the system. Nucleus and electron are all instances of the class `particle` defined below, starting from Line 79. The argument `part` of `baryc()` is a list of `particle` instances, comprising the nucleus and the two electrons. Each instance has its own variables, here we are interested in mass, `m`, coordinates, `x` and `y`, and velocity components, `vx` and `vy`. Line 60 evaluates the total mass of the atom, `mtot`. Note that the `for` loop inside the argument of `sum()` runs over all the elements `zz` of the list `part`, quantity `zz.m` being the mass of each particle. Lines 61–64 evaluate the position (`cx,cy`) and velocity components (`cvx,cvy`), of the barycenter. Function `SetBaryc()` moves nucleus and electrons to their positions in the barycenter reference frame, where the origin is located on the barycenter and the total momentum is zero.

```
78 # ................................................. Class particle
79 class particle:
80   def __init__(self,mass,charge,frict,x,y,vx,vy):
81     self.m=mass
82     self.q=charge
83     self.fr=frict
84     self.x=x
85     self.y=y
86     self.trailmin=np.sqrt(self.x**2+self.y**2)*0.05
87     self.vx=vx
88     self.vy=vy
89     if self.q>0:              # nucleus
90       self.col='red'
91       self.rad=8
92     else:                     # electron
93       self.col='blue'
94       self.rad=4
95     self.image=canvas.create_oval(Ox+int(scale*self.x-self.rad),\
96       int(Oy-scale*self.y+self.rad),int(Ox+scale*self.x+self.rad),\
97         int(Oy-scale*self.y-self.rad),fill=self.col,outline=self.col)
98     self.trail=[self.x,self.y]*TrailLength
99     self.ScaledTrail=[0.0,0.0]*TrailLength
100    self.TrailImg=canvas.create_line(self.ScaledTrail,fill=self.col)
101  # ................................................. move particle
102  def move(self):
103    canvas.coords(self.image,Ox+scale*self.x-self.rad,\
104      Oy-scale*self.y+self.rad,Ox+scale*self.x+self.rad,\
105        Oy-scale*self.y-self.rad)
106  def UpdateTrail(self):
107    if abs(self.x-self.trail[-2])+abs(self.y-self.trail[-1])>self.trailmin:
108      del self.trail[:2]
109      self.trail.append(self.x)
110      self.trail.append(self.y)
111  def DrawTrail(self):
112    self.ScaledTrail[::2]=[Ox+scale*zz for zz in self.trail[::2]]
113    self.ScaledTrail[1::2]=[Oy-scale*zz for zz in self.trail[1::2]]
114    canvas.coords(self.TrailImg,self.ScaledTrail)
```

Lines 79–114 define the class `particle`. Each instance of the class has its own values for mass, charge, linear drag coefficient (`frict`), position (`x`,`y`), and velocity (`vx`,`vy`), initialized by the `__init__()` method. The instance variable `trailmin`, equal to 0.05 times the distance of the particle from the origin, is the minimum distance between two consecutive points of the trail. If the charge of the particle is positive (nucleus) the particle is represented by a red circle of radius 8 px on the canvas. If the charge is negative (electron) the particle is represented by a blue circle of radius 4 px (Lines 89–94). Lines 95–97 draw the first image of the particle on the canvas. Line 98 initializes the list `trail`, comprising `TrailLength` couples of (x, y) coordinates that will describe the trail of the particle. Line 99 initializes the list `ScaledTrail`, whose elements equal the corresponding elements of `trail` multiplied by `scale`. Line 100 generates `TrailImg`, the first image of the trail on the canvas. Both `image` and `TrailImg` will be updated at each step (frame) of our animation.

Method `move()`, at lines 102–105, moves the particle image to its new position at the next animation step. This is done through the method `canvas.coords()`, whose arguments are the image to be moved, `self.image`, and the coordinates of the new position.

Method `UpdateTrail()` checks if the present position of the particle, (`self.x`,`self.y`), differs by more than `self.trailmin` from the last couple of coordinates of the list `self.trail`. If so, the coordinates of the first (oldest) point of the polyline are removed, and the coordinates of the new position are added at the end of the list.

Method `DrawTrail()` evaluates the coordinates of the polyline to be drawn on the canvas. Since we use usual Cartesian coordinates for our calculations, and prefer to have the origin at the center of the canvas, we convert our coordinates by the rules

$$x_{canv} = O_x + S x_{Cart} \ , \quad y_{canv} = O_y - S y_{Cart} \ , \tag{8.6}$$

where S is the scale factor `scale`, and we take into account that the canvas y axis is directed downwards. Since x and y coordinates alternate in the list, we use *slicing* for the conversion.

```
115 # ............................................ create input vector
116 def WriteInput(bodies,val,InpV,vect):
117   nn=7*len(bodies)
118   InpV[:nn:7]=[zz.m for zz in bodies]
119   InpV[1:nn:7]=[zz.q for zz in bodies]
120   InpV[2:nn:7]=[zz.fr for zz in bodies]
121   InpV[3:nn:7]=vect[::4]=[zz.x for zz in bodies]
122   InpV[4:nn:7]=vect[1::4]=[zz.y for zz in bodies]
123   InpV[5:nn:7]=vect[2::4]=[zz.vx for zz in bodies]
124   InpV[6:nn:7]=vect[3::4]=[zz.vy for zz in bodies]
125   InpV[nn::]=val
126 # ............................................. Read Entry values
127 def ReadInput(InpV,bodies,val,vect):
128   nn=7*len(bodies)
129   for i,zz in enumerate(bodies):
130     zz.m=InpV[i*7]
131     zz.q=InpV[i*7+1]
132     zz.fr=InpV[i*7+2]
133     zz.x=vect[i*4]=InpV[i*7+3]
134     zz.y=vect[i*4+1]=InpV[i*7+4]
135     zz.vx=vect[i*4+2]=InpV[i*7+5]
136     zz.vy=vect[i*4+3]=InpV[i*7+6]
137   val[::]=InpV[nn::]
```

Lines 116–125 define the function `WriteInput()`, which copies the instance variables of the single particles into the lists `InpV` and `vect`. List `InpV` is used for interactively changing parameters during the program execution, through the toolbar entries. List `vect` is the array of initial conditions for the numerical solution of the differential equations by `odeint()`. The name `vect` is used as local variable in the `WriteInput()` and `ReadInput()` functions, in the main program the corresponding list is called `y`, see Lines 178, 179 and 275. Line 125 adds the values of the list `values`, defined at Line 175 below, at the end of `Inpv`. The first $7n$ elements of `InpV` are the instance variables of the n particles ($n = 3$ in our case, but you can easily add new electrons between Lines 172 and 173), and its last 4 elements are the elements of `values`. The two lists are thus

$$\begin{aligned}&\texttt{InpV} = [m_0, q_0, \eta_0, x_0, y_0, \dot{x}_0, \dot{y}_0, m_1, q_1, \eta_1, x_1, y_1, \dot{x}_1, \dot{y}_1, m_2, q_2, \eta_2, x_2, y_2, \dot{x}_2, \dot{y}_2, k_e, dt, \Delta\tau, N_{\text{trail}}]\,,\\ &\texttt{vect} = [x_0, y_0, \dot{x}_0, \dot{y}_0, x_1, y_1, \dot{x}_1, \dot{y}_1, x_2, y_2, \dot{x}_2, \dot{y}_2]\,,\end{aligned} \tag{8.7}$$

where the numerical subscripts refer to the particles, and η_i stands for the linear drag-force coefficient on particle i. Note the use of list slicing in Lines 118–124. The elements of `InpV` appear in the entry windows of the toolbar, see Fig. 8.4.

Once we have modified some of the parameter values in the toolbar entries, function `ReadInput()`, defined at Lines 127–137, copies our new values into the class particle instances and into the `vect` and `values` lists.

```
138 # ................................................... vect2bodies
139 def vect2bodies(vect,bodies):
140   for i,zz in enumerate(bodies):
141     zz.x=vect[4*i]
142     zz.y=vect[4*i+1]
143     zz.vx=vect[4*i+2]
144     zz.vy=vect[4*i+3]
```

Function `vect2bodies()`, defined at Lines 139–144, copies the positions and velocities of the particles from the list `vect`, see (8.7), into the particle instance variables.

```
145 # ................................................... root window
146 root=Tk()
147 root.title('Classical Helium Atom')
148 root.bind('<Return>',ReadData)
149 root.bind('<Control-plus>',ScaleUp)
150 root.bind('<Control-minus>',ScaleDown)
151 # ........................................................ canvas
152 canvas=Canvas(root,width=cw,height=ch,background="#ffffff")
153 canvas.grid(row=0,column=0)
154 # ....................................................... toolbar
155 toolbar=Frame(root)
156 toolbar.grid(row=0,column=1, sticky=N)
```

Lines 146–150 create the root window and bind the <Return> keyboard key to the `ReadData()` function, the <Control-plus> key combination (simultaneous pressing of the <Ctrl> and <+> keys) to `ScaleUp()`, and <Control-minus> to `ScaleDown()`. A canvas and a toolbar are created in the root window.

```
157 # ....................................................... buttons
158 nr=0
159 StartButton=Button(toolbar,text="Start",command=StartStop,width=11)
160 StartButton.grid(row=nr,column=0,sticky=W)
```

```
161 AdjustButton=Button(toolbar, text="Set Barycenter",\
162   command=SetBaryc,width=11)
163 AdjustButton.grid(row=nr,column=1,sticky=W)
164 nr+=1
165 CloseButton=Button(toolbar, text="Exit", command=StopAll,width=11)
166 CloseButton.grid(row=nr,column=0,columnspan=2,sticky=W)
167 nr+=1
```

Three buttons, bound to the functions `StartStop()`, `SetBaryc()` and `StopAll()`, respectively, are created in the toolbar, located as in Fig. 8.4.

```
168 # ............................................... Create bodies
169 part=[]
170 part.append(particle(4.0*mp,2.0*q,0.0,0.0,0.0,0.0,0.0))
171 part.append(particle(me,-q,0.0,-r1,0.0,0.0,-v1))
172 part.append(particle(me,-q,0.0,r2,0.0,0.0,v2))
173 nP=len(part)
174 # ........................................ Parameter-value list
175 values=[ke,dt,cycle,TrailLength]
176 # ............................................... input vector
177 InpV=[0]*(7*nP+len(values))
178 y=[0]*4*nP
179 WriteInput(part,values,InpV,y)
```

The nucleus, first and second electron are created at lines 170–173 as instances of the class `particle`, members of the list `part`. The nucleus, `part[0]`, has mass $4m_\mathrm{p}$ and charge $2q$, each electron, `part[1]` and `part[2]`, has mass m_e and charge $-q$. The nucleus is located in the origin with initial zero velocity, while the electrons are located at $y_1 = y_2 = 0$, $x_1 = -r_1$, $x_2 = r_2$, with initial zero x velocity components $\dot{x}_1^0 = \dot{x}_2^0 = 0$, and y velocity components $\dot{y}_1^0$ and $\dot{y}_2^0$. All values of the x position components and of the y velocity components are slightly changed if one presses the *Set Barycenter* button on the toolbar, which performs the coordinate transformation to the barycentric frame.

Line 175 generates the list `values`, comprising the Coulomb constant k_e, the numerical-integration step $\mathrm{d}t$, the required time interval between two successive animation frames, and the number of vertices in the trails following the particles.

Lines 177–179 generate the lists `InpV` and `y`, and initialize them using the function `WriteInput()`.

```
180 # .................................................. Input list
181 InputStr=[]
182 for i in range(len(part)):
183   InputStr.append('m'+sub[i])
184   InputStr.append('q'+sub[i])
185   InputStr.append('\u03B7'+sub[i])  #    eta
186   InputStr.append('x'+sub[i])
187   InputStr.append('y'+sub[i])
188   InputStr.append('vx'+sub[i])
189   InputStr.append('vy'+sub[i])
190 InputStr.append('Ke')
191 InputStr.append('dt')
192 InputStr.append('Cycle/ms')
```

```
193 InputStr.append('Tail')
194 # .................................. Labels and Entries for particles
195 InputLab=[]
196 InputEntry=[]
197 for i,zz in enumerate(InputStr):
198   InputLab.append(Label(toolbar,text=zz,font=("Helvetica",11)))
199   InputLab[i].grid(row=nr,column=0)
200   InputEntry.append(Entry(toolbar,bd=3,width=12))
201   InputEntry[i].grid(row=nr,column=1)
202   InputEntry[i].insert(0,"{:.3e}".format(InpV[i]))
203   nr+=1
204 # ................................................... time label
205 CycleLab0=Label(toolbar,text="Period:",font=("Helvetica",11))
206 CycleLab0.grid(row=nr,column=0)
207 CycleLab=Label(toolbar,text="␣␣␣␣␣",font=("Helvetica",11))
208 CycleLab.grid(row=nr,column=1,sticky=W)
209 nr+=1
210 # ................................................... scale label
211 ScaleLab0=Label(toolbar,text="Scale:",font=("Helvetica",11))
212 ScaleLab0.grid(row=nr,column=0)
213 ScaleLab=Label(toolbar,text="%10.3e"%(scale),font=("Helvetica",11))
214 ScaleLab.grid(row=nr,column=1,sticky=W)
215 nr+=1
```

Lines 181–193 generate the label strings for the entries in the toolbar, as seen in the left toolbar column of Fig. 8.4, `\u03B7` is the utf-8 encoding of the Greek letter η, denoting the linear-drag constant. Lines 195–203 create the labels and entries for modifying the program parameters,located in the toolbar. Lines 205–208 create two side-by-side labels in the toolbar, one with the word *Period:*, the other reporting the actual average time interval between consecutive frames during the program execution. Lines 211–214 create two labels reporting the value of the variable `scale`.

```
216 # ........................................................ function
217 def dfdt(yInp,t,pp):
218   nn=7*(len(yInp)//4)
219   mm=np.array(pp[:nn:7])                        # masses from InpV
220   qq=pp[1:nn:7]                                # charges from InpV
221   fr=np.array(pp[2:nn:7])                        # drags from InpV
222   # ..............................................................
223   x=yInp[::4]
224   y=yInp[1::4]
225   vx=np.array(yInp[2::4])
226   vy=np.array(yInp[3::4])
227   # ........................... Coulomb contribution to force
228   distx=x-(np.tile(x,(len(x),1))).T
229   disty=y-(np.tile(y,(len(y),1))).T
230   alpha=np.arctan2(disty,distx)
231   r2=np.square(distx)+np.square(disty)
232   np.fill_diagonal(r2,1.0)
233   q2=-ke*(np.tile(qq,(len(qq),1)).T*qq)
234   ff=np.divide(q2,r2)
235   np.fill_diagonal(ff,0.0)
236   fx=ff*np.cos(alpha)
```

```
237     fy=ff*np.sin(alpha)
238     # .................... Accelerations, including linear drags
239     ax=(fx.sum(axis=1)-(vx*fr))/mm
240     ay=(fy.sum(axis=1)-(vy*fr))/mm
241     # ...................................... Build output list
242     derivs=[0]*len(yInp)
243     derivs[::4]=vx
244     derivs[1::4]=vy
245     derivs[2::4]=ax
246     derivs[3::4]=ay
247     # ........................................................
248     return derivs
```

Lines 217–247 define the function `dfdt()`, called by `odeint()` at line 274. Argument `yInp` is the list $[x_0, y_0, \dot{x}_0, \dot{y}_0, x_1, y_1, \dot{x}_1, \dot{y}_1, \ldots]$ containing the positions and velocities of the particles at the beginning of each iteration step. Argument `t` is the list $[0, \mathrm{d}t]$ defined at line 249, comprising the time points at which positions and velocities must be evaluated by `odeint()`. Here the two elements of `t` simply determine the time length of the integration step. The function call at line 274 will pass the list `InpV`, containing all program parameters, as argument `pp`. Function `dfdt()` returns the list of the derivatives of the elements of `yInp`, namely $[\dot{x}_0, \dot{y}_0, \ddot{x}_0, \ddot{y}_0, \dot{x}_1, \dot{y}_1, \ddot{x}_1, \ddot{y}_1, \ldots]$.

Line 218 evaluates `nn`, the number of `pp` elements referring to the particles (the last four elements of `pp` being the elements of `values`). The number of particles N_p equals the length of `yInp` divided by 4, since each particle has its own x and y components of position and velocity. Thus `nn` equals N_p multiplies by 7, since each particle has its own mass, charge and linear-drag coefficients, plus the x and y components of its initial position and initial velocity. Lines 219–221 create the lists `mm`, `qq` and `fr`, comprising the particle masses, charges, and linear-drag coefficients, respectively. These lists are obtained by appropriately slicing List `pp`, as discussed in Sect. 1.13. Lines 223–226 obtain the four lists of positions and velocities at the beginning of the integration step by slicing the list `yInp`.

Lines 228–237 evaluate the Coulomb forces acting on particles. Lines 228 and 229 evaluate the matrices of the x and y components of the interparticle distances. According to the array operations discussed in Sect. 1.17, we have

$$\mathtt{distx} = \begin{pmatrix} 0 & x_1 - x_0 & x_2 - x_0 \\ x_0 - x_1 & 0 & x_2 - x_1 \\ x_0 - x_2 & x_1 - x_2 & 0 \end{pmatrix}, \qquad \mathtt{disty} = \begin{pmatrix} 0 & y_1 - y_0 & y_2 - y_0 \\ y_0 - y_1 & 0 & y_2 - y_1 \\ y_0 - y_2 & y_1 - y_2 & 0 \end{pmatrix}. \tag{8.8}$$

Note that here the Python matrix-manipulation routines perform some unnecessary calculations because of the skew-symmetry of both matrices. Actually the evaluation of a single triangular submatrix, excluding the diagonal elements, would be sufficient for each matrix. However the available Python matrix-manipulation routines are pre-compiled and optimized, thus they run much faster than user-written nested loops evaluating the triangular submatrices only. The i, j elements of the upper triangular submatrix of matrix `alpha`, evaluated at Line 230, are thus the α_{ij} angles shown in

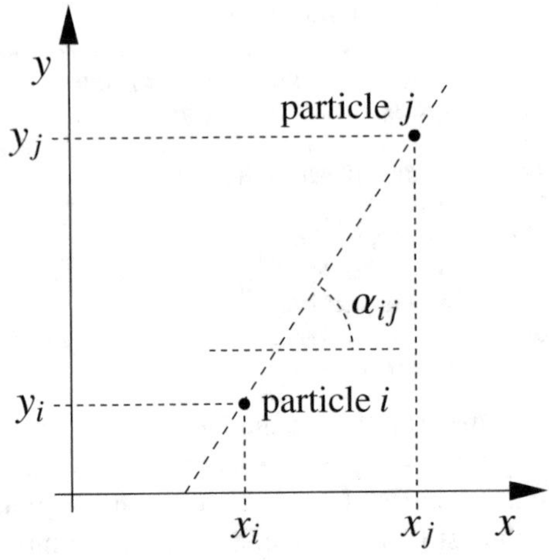

Fig. 8.5 Particles i and j, and the angle α_{ij}

Fig. 8.5. The i, j elements of the lower triangular submatrix of `alpha` equal $\alpha_{ji} + \pi$, therefore $\sin\alpha_{ij} = -\sin\alpha_{ji}$ and $\cos\alpha_{ij} = -\cos\alpha_{ji}$. Again here and in the following, Python matrix-manipulation routines perform some unnecessary, but fast, calculations. On the diagonal, where both arguments of function `numpy.arctan2()` are zero, the function can return either zero or π, but this is irrelevant for our purposes, since we don't need the diagonal elements of `alpha`. Line 231 creates the matrix `r2` whose elements r_{ij}^2 are the squares of the distances between particle i and particle j. Line 232 arbitrarily fills the diagonal of `r2` with ones, in order to avoid divisions by zero on the matrix diagonal at Line 234, since $r_{ii}^2 = 0$. The resulting "artificial" diagonal elements are not used in the subsequent calculations. Line 233 creates the matrix `q2`, with elements $(\mathtt{q2})_{ij} = -k_e q_i q_j$. Line 234 obtains the matrix `ff` by Hadamard (element by element) matrix division. The off-diagonal elements of `ff` are thus

$$-k_e \frac{q_i q_j}{r_{ij}^2} \tag{8.9}$$

corresponding to the magnitudes the Coulomb forces between the particle pairs i, j, with a plus sign if the force is attractive, and a negative sign if the force is repulsive. The diagonal elements of `ff` are zeroed by Line 235. Lines 236 and 237 form the matrices of the x and y components of the inter-particle forces. Thus, the sum of the elements of row i of `fx` (`fy`) is the x (y) component of the total Coulomb force acting on particle i. These sums of the row elements are performed at Lines 239–240, where also the contribution of the linear-drag force is added, and the total force components are divided by the particle masses in order to obtain the lists `ax` and `ay` of the particle acceleration components.

Lines 242–246 build the list `derivs` comprising the derivatives of the elements of `yInp`, which function `dfdt()` returns at Line 248, the last line of the `dfdt()` definition code.

```
249 # ........................................ numerical time interval
250 t=[0.0,dt]
251 tt0=time.time()
252 tcount=0
253 # .......................................... draw coordinate axes
254 canvas.create_line(0,Oy,cw,Oy, fill="black")
255 canvas.create_line(Ox,0,Ox,ch, fill="black")
256 # ................................. Create barycenter image on canvas
257 bc=canvas.create_oval(Ox-bcrad,Oy-bcrad,Ox+bcrad,Oy+bcrad, fill="black")
```

Line 250 builds the list `t` of the time points at which `odeint()` evaluates the solutions of the differential equations at each animation step. Line 251 stores the seconds elapsed since the *epoch* in variable `tt0`. Line 252 sets the iteration counter `tcount` to zero. Lines 254 and 255 draw the x and y axes on the canvas, and Line 257 draws a first image of the system barycenter.

```
258 # ..........................................................
259 while RunAll:
260     StartIter=time.time()
261     # ................................................... draw bodies
262     for zz in part:
263         zz.move()
264         zz.UpdateTrail()
265         zz.DrawTrail()
266     # ................................................ center of mass
267     cx,cy=baryc(part)[0]
268     cx*=scale
269     cy*=scale
270     canvas.coords(bc,Ox+cx-bcrad,Oy-cy-bcrad,Ox+cx+bcrad,Oy-cy+bcrad)
271     canvas.update()
```

Lines 259–315 comprise our main animation loop. Line 260 stores the seconds elapsed from *epoch* to the beginning of each iteration in `StartIter`. Loop 262–265 moves all particles to there new positions, and updates the particle trails. Lines 267–270 evaluate the new position of the barycenter and draw its image on the canvas. Line 271 updates the canvas.

```
272     # ......................................................... motion
273     if RunIter:
274         # ................................................... next step
275         psoln = odeint(dfdt,y,t,args=(InpV,))
276         y=psoln[1,:]
277         vect2bodies(y,part)
```

If `RunIter` is *True* lines 275–277 call `odeint()` to solve the differential equations for the particle motions and update the list `y`, comprising the particle positions and velocities at the next step. Line 277 copies the new positions and velocities stored in `y` into the class particle instances, so that the class methods `move()`, `UpdateTrail()` and `DrawTrail` can update the animation frame at the next iteration step.

```
278 # ..........................................................
279 else:
280   if NewBaryc:
281     ReWrite=True
282     WriteInput(part,values,InpV,y)
283     NewBaryc=False
284   elif GetData:
285     for i,zz in enumerate(InputEntry):
286       try:
287         InpV[i]=float(zz.get())
288       except ValueError:
289         pass
290     ReWrite=True
291     GetData=False
```

Lines 280–304 are executed only if `RunIter` is *False*. If `NewBaryc` is *True*, i.e., if the toolbar button <Set Barycenter> has been pressed, thus calling function `SetBaryc()`, variable `ReWrite` is set to *True*, and lists `InpV` and `y` are updated with the new initial positions and velocities in the barycenter reference frame. Eventually, `NewBaryc` is set to *False*.

If `GetData` is *True*, i.e., if the <Return> key has been pressed, Loop 285–289 rereads all values in the toolbar entries and updates list `InpV`. Variable `ReWrite` is set to *True* and `GetData` to *False*.

```
292   if ReWrite:
293     ReadInput(InpV,part,values,y)
294     for zz,yy in zip(InputEntry,InpV):
295       zz.delete(0,'end')
296       zz.insert(0,"{:.3e}".format(yy))
297     dt=values[1]
298     t=[0.0,dt]
299     cycle=int(values[2])
300     TrailLength=int(values[3])
301     for zz in part:
302       zz.trail=[zz.x,zz.y]*TrailLength
303       zz.ScaledTrail=[0.0,0.0]*TrailLength
304     ReWrite=False
```

If `ReWrite` is *True*, i.e., if either `NewBaryc` or `GetData` was *True*, the values stored in List `InpV` are copied into the class-particle instances, and into the lists `values` and `y`. Loop 294–296 rewrites the values in the toolbar entry-windows formatted with 3 digits after the decimal point. Lines 297–298 update variable `dt` and List `t`. Lines 299–300 update the integer values of `cycle` (in ms) and `TrailLength`. Loop 301–303 reinitializes the particle trails with the new lengths. Eventually, `ReWrite` is set to *False*.

```
305 # ............................................ cycle duration
306 tcount+=1
307 if tcount==10:
308   tcount=0
309   ttt=time.time()
310   elapsed=ttt-tt0
```

```
311         CycleLab['text']="%8.3f"%(elapsed*100.0)+" ms"
312         tt0=ttt
313       ElapsIter=int((time.time()-StartIter)*1000.0)
314       canvas.after(cycle-ElapsIter)
```

Line 306 increases the iteration counter `tcount`. Lines 307–312 are executed every 10 iterations. Variable `elapsed` is the number of seconds elapsed during the last 10 iterations, which, multiplied by 100, gives the average duration of each iteration in ms. The value is written in the last-but-one label at the bottom of the toolbar. Variable `tt0` is reinitialized.

Variable `ElapsIter` at Line 313 is the time elapsed since the start of the present iteration, in ms. Line 314 delays the start of the next iteration till `cycle` ms have elapsed since the beginning of the present iteration. If `ElapsIter>cycle`, the argument of `canvas.after()` is negative, and Line 314 is simply ignored.

```
315   #------------------------------------------------------------
316   root.destroy()
```

It is interesting to experiment on this script, using the entry labels in the toolbar to change the initial parameter values, and observe how the behavior of our "classical atom" is modified. If the initial positions and velocities of the electrons are modified, in most cases the behavior becomes unstable, and, eventually, "autoionization" may result. Changing the sign of the velocity of either electron is sufficient to observe such instabilities.

A further possibility is adding new electrons between Lines 173 and 174, with arbitrary initial positions and velocities.

8.3.1 A "C-Style" `dfdt()` Alternative Function

Function `dfdt()`, defined at Lines 217–248, can be replaced by the following code, where the nested loops at Lines 230–246 avoid the unnecessary calculation of the whole skew-symmetric matrices. This code can be easier to read for the C-minded programmer, but executes more slowly.

Listing 8.4 C-style

```
217   #.................................................. function
218   def dfdt(yInp,t,pp):
219     nn=7*(len(yInp)//4)
220     mm=pp[:nn:7]                      # masses from InpV
221     qq=pp[1:nn:7]                     # charges from InpV
222     fr=pp[2:nn:7]                     # drags from InpV
223     #.............................................................
224     x=yInp[::4]
225     y=yInp[1::4]
226     vx=yInp[2::4]
227     vy=yInp[3::4]
228     Fx=list(-array(vx)*array(fr))   # drag contribution to force x
229     Fy=list(-array(vy)*array(fr))   # drag contribution to force y
```

```
# ........................... Coulomb contribution to force
i=1
while i<nP:
  j=0
  while j<i:
    deltax=x[i]-x[j]
    deltay=y[i]-y[j]
    r2=deltax**2+deltay**2
    alpha=arctan2(deltay,deltax)
    ff=-ke*qq[i]*qq[j]/r2
    fx=ff*cos(alpha)
    fy=ff*sin(alpha)
    Fx[i]-=fx
    Fx[j]+=fx
    Fy[i]-=fy
    Fy[j]+=fy
    j+=1
  i+=1
# ..............................................................
derivs=[0]*len(yInp)
derivs[::4]=vx
derivs[1::4]=vy
derivs[2::4]=list(array(Fx)/array(mm))
derivs[3::4]=list(array(Fy)/array(mm))
# ..............................................................
return derivs
```

8.4 Exercises

Exercise 8.1 Six astronomical bodies have masses $m_1 = 1.3 \times 10^{22}$ kg, $m_2 = 1.6 \times 10^{21}$ kg, $m_3 = 7.5 \times 10^{14}$ kg, $m_4 = 5.0 \times 10^{16}$ kg, $m_5 = 1.6 \times 10^{16}$ kg, and $m_6 = 5.0 \times 10^{16}$ kg, respectively. At time $t = 0$ they are aligned on the x axis of a Cartesian reference frame with its origin at the barycenter of the six bodies, as shown in Fig. 8.6. The coordinates of the six bodies are $x_1 = -2.137 \times 10^6$ m, $x_2 = 1.736 \times 10^7$ m, $x_3 = 4.266 \times 10^7$ m, $x_4 = 4.869 \times 10^7$ m, $x_5 = 5.778 \times 10^7$ m, and $x_6 = 6.486 \times 10^7$ m. All six velocities are directed along the y axis, with values $v_{y1} = -19.4$ m/s, $v_{y2} = 158$ m/s, $v_{y3} = 151$ m/s, $v_{y4} = 142$ m/s, $v_{y5} = 130$ m/s, and $v_{y6} = 123.6$ m/s, respectively (this system is a very rough approximation of Pluto and its five known moons: Charon, Styx, Nix, Kerberos and Hydra. The probability of the assumed initial alignment is obviously zero!) Following Script 8.3, write a program simulating the subsequent motion of the six bodies under the action of the mutual gravitational forces. The gravitational constant is $G = 6.67408 \times 10^{-11}$ $\text{m}^3\,\text{kg}^{-1}\,\text{s}^{-2}$. Hints: (i) create a class to which the six bodies belong; (ii) use a scale of 5×10^{-6} px/m; (iii) use a "real" time interval between two successive frames $\Delta t = 600$ s.

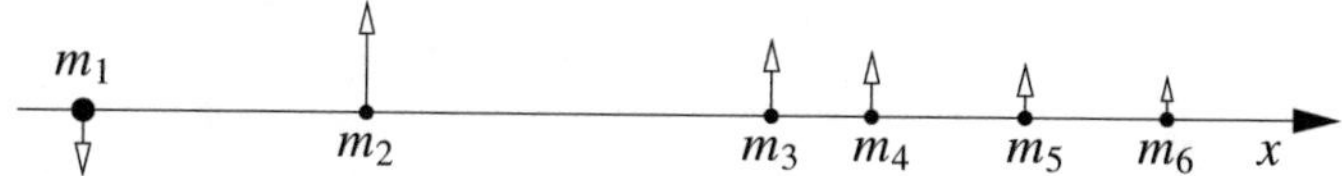

Fig. 8.6 The six bodies at $t = 0$

Exercise 8.2 If you let Script 8.5, which answers Exercise 8.1, run long enough (30 min or longer) you observe that the system is slowly shifting upwards in the canvas. Why? How can you correct this behavior?

Chapter 9
3D Graphics and Animation with VPython

9.1 Introduction and Installation

VPython comprises the Python programming language plus a 3D graphics module called `vpython`. VPython allows users to create objects such as spheres, cones, ...in three-dimensional space and displays these objects in a window of the browser. This makes it easy to create simple visualizations, allowing programmers to focus more on the computational aspect of their programs. VPython is used for various purposes, and thanks to its simplicity it is even used by students who are new to computer programming. VPython 3D animations use WebGL ("Web Graphics Library", a JavaScript application programming interface), therefore they need a browser to run. Most current browsers support WebGL, however this is not true for versions of Internet Explorer older than version 11.

Since VPython is a very powerful and sophisticated software, covering it completely would require a dedicated book. Our aim in the context of the present book is to give you just a basic introduction. This should enable you to write with confidence your first programs. Once you have reached this level, it will be easy for you to search the internet, the VPython and `glowscript` web sites in particular, when you need more sophisticated techniques. On these sites you can find many examples of programs of physical interest, as well as a lot of documentation.

An advantage of VPython is that, if you wish, you don't need to install software in your computer, not even Python, to run it. All you need is going to the `glowscript.org` web site and create your free account there. The Help of the site will provide you full documentation and example programs, in addition to the possibility of writing your own VPython programs on the site. You can also ask the site to load local copies of the `glowscript` libraries and a program editor into your

---

Electronic supplementary material The online version of this chapter (https://doi.org/10.1007/978-3-030-45027-4_9) contains supplementary material, which is available to authorized users.

G. Moruzzi, *Essential Python for the Physicist*,
https://doi.org/10.1007/978-3-030-45027-4_9

computer. This will allow you to run VPython offline even if you have not previously installed Python. There is only one caveat: when using `glowscript` VPython, whether online or offline, one does not have access to the world of Python modules, whereas one does have such access when using the vpython module with installed Python. Here we assume that you run your programs with Python and VPython installed in your computer, however, all the scripts of this chapter can be used also on the glowscript site.

If you have installed the Anaconda Python distribution with Python version 3.5.3 or later (see Sect. 1.2.1) you can install the VPpython module *VPython 7* by typing in a terminal

```
$> conda install -c vpython vpython
```

otherwise, if you are using Python from a terminal, you can install VPython from the command line by typing

```
$> pip3 install vpython
```

9.2 The VPython Vector Object

The first topic we consider here is VPython's *vector object*, even if it is not a displayable object. However the vector object is a powerful aid to the computations involved in 3D graphics and animations. Its properties are similar to vectors used in science and engineering. Actually, the Vpython vector object can be useful in your programs even if you are not using VPython graphics and animation. You can learn how the vector object works by doing some experimenting in interactive mode. In the following listing, as usual, lines are numbered only for reference in the text.

```
>>> from vpython import *
>>> a=vector(2,3,4)
>>> b=vec(2.5,2.5,1)
>>> a
<2, 3, 4>
>>> a+b
<4.5, 5.5, 5>
>>> a-b
<-0.5, 0.5, 3>
>>> dot(a,b)
16.5
>>> cross(a,b)
<-7, 8, -2.5>
>>> dot(a,a)
29.0
>>> cross(a,a)
<0, 0, 0>
>>> 2.5*a
<5, 7.5, 10>
```

Line 1 imports everything from VPython. If you use VPython graphics and animation, you will not need Tkinter, and VPython will be the only package from which you import everything. It is not mandatory, but strongly recommended, not to use the command "`import *`" for other packages. Line 2 defines the variable `a` as a vector of components ($a_x = 2, a_y = 3, a_z = 4$), line 3 defines the vector `b` analogously, using the abbreviation `vec()` for `vector()`. The two forms `vector()` and `vec()` are completely equivalent. Being in interactive mode, line 4 asks to print the vector `a`, which is printed at Line 5. Lines 6–9 print the sum and difference of the two vectors. Note that the vector components are written within angle brackets. Line 10 prints the dot product (scalar product) of `a` and `b`, in mathematical notation $\boldsymbol{a} \cdot \boldsymbol{b}$. Line 12 prints the cross product of the two vectors, $\boldsymbol{a} \times \boldsymbol{b}$. Lines 14–17 print the dot and cross products of `a` by itself. Line 18 asks Python to print the product of the scalar 2.5 by the vector `a`.

```
20 >>> mag(a)
21 5.385164807134504
22 >>> sqrt(dot(a,a))
23 5.385164807134504
24 >>> a.mag
25 5.385164807134504
26 >>> mag2(a)
27 29.0
28 >>> hat(a)
29 <0.371391, 0.557086, 0.742781>
30 >>> mag2(hat(a))
31 1.0
```

Line 20 prints the magnitude of `a`, or $|\boldsymbol{a}|$. Typing `a.mag` is equivalent to typing `mag(a)`. Line 26 prints the vector magnitude squared, or $|\boldsymbol{a}|^2 = \boldsymbol{a} \cdot \boldsymbol{a}$. Line 28 prints a unit vector in the direction of `a`, in mathematical notation $\hat{\boldsymbol{a}} = \boldsymbol{a}/|\boldsymbol{a}|$. For convenience, `hat(vector(0,0,0))` is defined as `vector(0,0,0)`. Valid alternative notations are `a.mag2` and `a.hat`, respectively.

```
32 >>> c=rotate(a,angle=pi,axis=vector(0,1,0))
33 >>> c
34 <-2, 3, -4>
35 >>> d=rotate(a,angle=pi/2,axis=vector(0,0,1))
36 >>> d
37 <-3, 2, 4>
```

Line 32 defines the vector `c`, obtained by rotating the vector `a` by π radians around the y axis. Note that VPython provides the value of π as `pi`, and you don't need to import it from the `numpy` or `math` packages. Line 35 defines the vector `d`, obtained by rotating the the vector `a` by $\pi/2$ radians around the z axis. Note that the magnitude of the vector used as rotation axis is irrelevant: only its direction matters.

```
38 >>> diff_angle(a,b)
39 0.5846453917216585
40 >>> degrees(0.5846453917216585)
41 33.49771345742379
42 >>> radians(33.49771345742379)
```

```
43 0.5846453917216585
44 >>> a.hat=b
45 >>> a
46 <3.66414, 3.66414, 1.46566>
47 >>> a.mag2
48 29.0
49 >>> diff_angle(a,b)
50 0.0
```

Line 38 evaluates the angle, in radians, between the vectors `a` and `b`, Line 40 converts the angle to degrees, Line 42 reconverts the angle back to radians. Line 44 changes the direction of vector `a` to that of `b`, leaving the magnitude of `a` unchanged.

9.3 VPython Displayable Objects

9.3.1 Sphere

As a first example, you can enter interactive Python and type the following two commands

Listing 9.1 Sphere
```
1 >>> from vpython import *
2 >>> sphere()
```

Line 1 imports everything from VPython, as already discussed for Line 1 of the code of Sect. 9.2. As soon as you type Line 2, VPython opens a 3D canvas, named `scene`, in a browser window. The canvas is shown in Fig. 9.1. If you have a browser already running in your computer, the canvas will appear in its window. Otherwise, VPython will first launch a browser available in your computer, then open its canvas in it. The canvas has a black background and displays a 3D white sphere. The command `sphere()` at Line 2 actually tells many things to VPython. First, it tells that you need a 3D canvas where to create 3D objects, and, in particular that you want to create a sphere in it. It also tells that you need a camera pointing at the scene from an appropriate distance such that the image is well-placed in the scene, that you need some default lighting so that you can see the objects in the scene, and that you need mouse controls to let you zoom and rotate and pan the camera. All these requirements (position, radius and color of the sphere, camera position, light positions, ...) are given default values in the present case. We shall see below how you can impose values of your choice instead of the defaults. But it is important to note that you don't have to specify all of the details in order to get something to display. The default background color of the canvas is black (`scene.background=vector(0,0,0)`, equivalent to `scene.background=color.black`), as seen in Fig. 9.1. You can change this by adding, for instance, the following line to Listing 9.1. See Sect. G.1 of the Appendix.

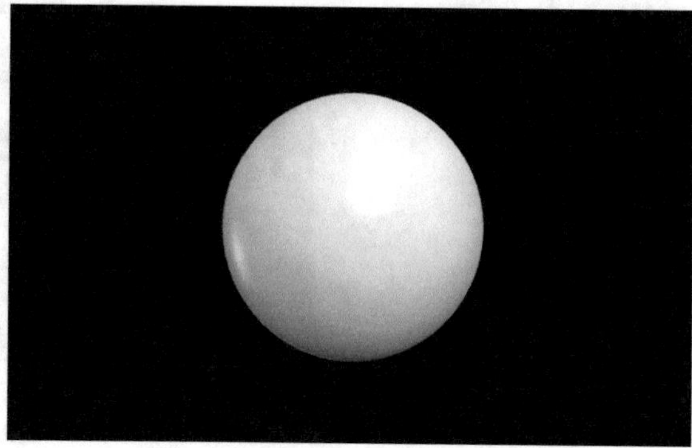

Fig. 9.1 VPython 3D canvas showing a sphere

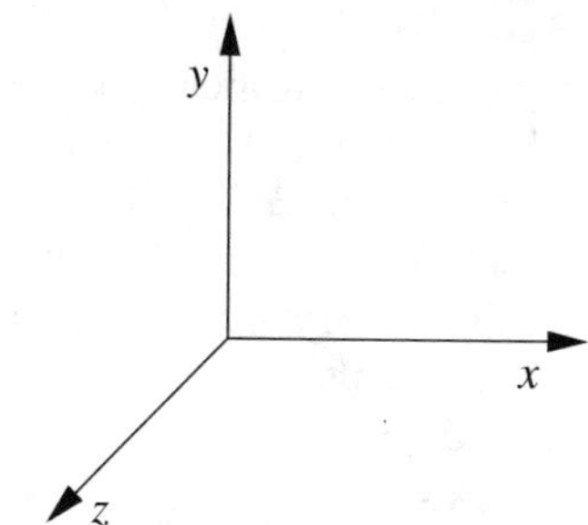

Fig. 9.2 Cartesian axes of the VPython canvas

```
>>> scene.background=color.blue
```

Objects are drawn in the canvas and positioned according to a Cartesian coordinate system with the x axis horizontal and directed toward the right, the y axis directed upwards and the z axis directed toward you, out of the computer monitor, as shown in Fig. 9.2. The default value for the sphere position is `pos=vector(0,0,0)`, meaning that the center of the sphere is located at the origin of the coordinates ($x = 0$, $y = 0$, $z = 0$). The default value of the sphere radius is `radius=1`. All lengths and positions are in arbitrary units which, however, must be consistent with one another. VPython automatically scales the image. Automatic scaling occurs dynamically during a program execution. You can prevent this with the command `scene.autoscale=False`, see Exercise 9.2. If you change Listing 9.1 to

```
>>> from vpython import *
>>> sphere(pos=vec(2,0,0))
```

you will observe the sphere located at the right of the canvas.

By default, the scene is illuminated by two distant lights (distant means that the light rays are parallel) defined by

```
distant_light(direction=vector(0.22, 0.44,0.88),color=color.gray(0.8))
distant_light(direction=vector(-0.88,-0.22,-0.44),color=color.gray(0.3))
```

This means that the light sources are located at two far-away points in the directions of the two unit vectors (0.22, 0.44, 0.88) and (−0.88, −0.22, −0.44), and the light

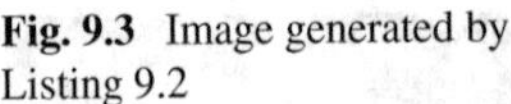

Fig. 9.3 Image generated by Listing 9.2

propagates in direction opposite to the unit vectors. Maximum light intensity is coded `color.gray(1)`, while minimum light intensity is `color.gray(0)` You can get rid of these default lights by typing the command scene.lights = [], which sets the light list to empty. This will make the canvas very dark, and the image almost invisible. Then you can add your local or distant light sources as discussed in Sect. 9.4.

You can also give the sphere a radius and a color of your choice by typing, for instance

Listing 9.2 ColorBall.py
```
1 >>> from vpython import *
2 >>> scene.background=vec(0,0,0.8)
3 >>> ball=sphere(pos=vec(1,0,0),radius=1.5,color=vec(0.9,0.9,0.2))
```

where you have moved the center of the sphere from (0, 0, 0) to (1, 0, 0). Writing `ball=sphere(...)` you create an object named `ball`, whose attributes can later be changed by further commands. For instance, you can change its position by typing `ball.pos=vec(new_x,new_y,new_z)`. We shall see that this is very important for creating animations. In Line 3 we have also set the sphere radius equal to 1.5 and the sphere color to yellow, as shown in Fig. 9.3.

The three "vector" components of the color are the usual red, green and blue intensities of the RGB additive color model. Obviously a color is not a vector, but the primary colors of the RGB model, which we met in Sect. 6.2, constitute a triad like the three components of an ordinary vector. VPython finds it convenient to use its vector object for managing RGB color definitions. The RGB color components are expressed by real numbers in the closed intervals $0 \leq r, g, b \leq 1$, rather than by the hexadecimal values that we met in Section 6.2 for Tkinter. Here the command `color=vector(0.9,0.9,0.2)` sets the red and green intensities to 90% of their possible maximum value, and the blue intensity to 20% of its maximum possible value. This is, with very good approximation, equivalent to the TKinter hexadecimal code `#E5E533`.

Other possible attributes for the sphere are

- **opacity**, opacity of the object. The default value is 1, zero means that the object is completely transparent.

- **shininess**, light reflectivity of the object. Values must be in the closed interval [0, 1], the default value is 0.6.
- **emissive**, can be *True* or *False*. If *True*, local and distant lights are ignored, and the brightness is governed by the object's own color. An example of its use is to put an emissive sphere at the location of a `local_light`, which looks like a glowing lamp. The default for `emissive` is False.
- **size**. Default is `vector(2,2,2)`. Instead of specifying the radius, you can set

  ```
  size=vector(length,height,width)
  ```

  which means that the cross section of the sphere can be elliptical, making it like the ellipsoid object.

It is interesting to observe what happens if you turn the mouse scroll wheel while the mouse cursor is on the canvas, or if you move the mouse cursor on the canvas while (i) keeping the shift key of the keyboard pressed, or (ii) keeping the right mouse button pressed. This will be explained in Sect. 9.4.

9.3.2 Box

Listing 9.3 adds a *rectangular cuboid* centered at $(-10, -3, 0)$, with edges parallel to the coordinate axes, to a white sphere located at $(2.5, 2.5, 4)$, as shown in Fig. 9.4. Lines 3 and 4 set the canvas width and height in px, while Line 5 makes the canvas background white. Line 7 creates the rectangular cuboid (a *box*) centered at $(x = -10, y = -3, z = 0)$, of length (parallel to the x axis) 12, height (parallel to the y axis) 7.5, and width (parallel to the z axis) 8. Length, height and width of the box can also be specified as separate attributes, as `length=12,height=7.5, width=8`. You are invited to observe how the image changes if you rotate the camera by moving the mouse while keeping the right button pressed.

Listing 9.3 SphereCuboid.py

```
1 #!/usr/bin/env python3
```

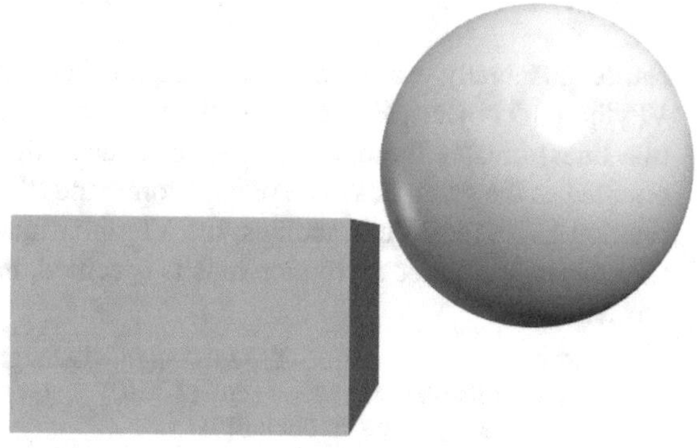

Fig. 9.4 VPython box and sphere

Fig. 9.5 Orienting and rotating boxes

```
2 from vpython import *
3 scene.width=800
4 scene.height=600
5 scene.background=vector(1,1,1)
6 ball=sphere(pos=vector(2.5,2.5,4),radius=5.4)
7 cuboid=box(pos=vector(-10,-3,0),size=vector(12,7.5,8),color=color.green)
```

The `box` object has the following attributes and default values:

- **`pos`** The coordinates of the center of the box. Default is `pos=vector(0,0,0)`.
- **`length`** In the x direction in the simple case, default is 1.
- **`height`** In the y direction in the simple case, default is 1.
- **`width`** In the z direction in the simple case, default is 1.
- **`size`** Combines (length, height, width). Default is `vector(1,1,1)`
- **`color`** Default is `vector(1,1,1)`, corresponding to white.
- **`shininess`**, light reflectivity of the object. Its value must be in the closed interval [0, 1], the default value is 0.6. Analogous to the `sphere` object.
- **`emissive`**, can be *True* or *False*. Analogous to the `sphere` object.
- **`axis`** The *axis* attribute gives a direction for the length of the box, so that the box is no longer aligned to the canvas coordinates. Now the length of the box is measured along the `axis`, rather than along x. For instance, the line

  ```
  cuboid=box(pos=vector(-25,-3,0),length=12,height=7.5,width=8,\
    color=color.green,axis=vector(2,1,0))
  ```

  generates `cuboid`, a box of the same dimensions of the one generated by Line 7 of Listing 9.3, but with its axis, and thus its length, aligned along the (2, 1, 0) vector, or the $(2/\sqrt{5}, 1/\sqrt{5}, 0)$ unit vector, as the green cuboid at the left of Fig. 9.5.

- **rotate()**. Actually, `rotate()` is not an attribute, but a method which works on VPython 3D objects. For instance, the first of the following commands (the first two lines) creates the object `cuboid1`, the yellow box of Fig. 9.5, originally oriented as the green box. The second command (third line) rotates it by 20°. The rotation angle must be in radians, the VPython function `radians()` performs the conversion. Since no rotaion axis is specified, the rotation occurs around the box axis.

  ```
  cuboid1=box(pos=vector(-7,-3,0),length=12,height=7.5,width=8,\
    color=color.yellow,axis=vector(2,1,0))
  cuboid1.rotate(angle=radians(20))
  ```

A different rotation axis can be specified. For instance, the following code first creates `cuboid2`, a red box originally oriented as `cuboid`, then rotates it by 45° around the y axis.

```
cuboid2=box(pos=vector(10,-3,0),length=12,height=7.5,width=8,\
  color=color.red,axis=vector(2,1,0))
cuboid2.rotate(angle=radians(45),axis=vector(0,1,0))
```

The following code generates the blue box and rotates it around the z axis.

```
cuboid3=box(pos=vector(25,-3,0),length=12,height=7.5,width=8,\
  color=color.blue,axis=vector(2,1,0))
cuboid3.rotate(angle=radians(45),axis=vector(0,0,1))
```

Rotations occur around an axis, relative to an origin. If the origin is not given, as in the above examples, the rotation is about the `pos` attribute of the object. The origin is specified by the `origin` attribute, for instance

```
cuboid.rotate(angle=radians(90),axis=vector(1,0,2),origin=vector(2,2,3))
```

9.3.3 Cylinder

Most of the cylinder attributes are analogous to the box attributes of the same name. An obvious difference is that the triad of attributes {`length`, `height`, `width`} is replaced by the pair {`length`, `radius`}. However, the attribute `size=vector(length, height, width)` can be specified also for a cylinder, generating a cylinder of elliptical, rather than circular, section. Another important difference is that, while the attribute `pos` refers to the coordinates of the center for a box and a sphere, it refers to the center of one base of the cylinder. This is clearly seen in Fig. 9.6, where the cube and the cylinder have the same `pos` attribute. The green cube is semitransparent, this is obtained by specifying `opacity=0.3`. The two commands generating Fig. 9.6 are

```
rod=cylinder(pos=vector(0,0,0),size=vector(4,2,1),color=vector(0.8,0.8,0.2))
cube=box(pos=vector(0,0,0),size=vector(3,3,3),color=color.green,opacity=0.3)
```

This is a short discussion of the cylinder attribute

- **`axis`** The axis points from `pos` to the other end of the cylinder, default is `axis=vector(1,0,0)`. Setting the axis makes length equal to the magnitude of the axis. An object's `axis` and `up` attributes are always perpendicular to each other. Changing the direction of axis also changes the direction of `up` so that the two directions always remain at right angles to each other.
- **`up`** Which side of the cylinder is "up"; this has only a subtle effect on the 3D appearance of the cylinder unless a non-smooth texture is specified or the cross section is oval; the default is `up=vector(0,1,0)`. An object's axis and up attributes are always perpendicular to each other. Changing the direction of `up` also changes the direction of axis so that the two directions always remain at right angles to each other.

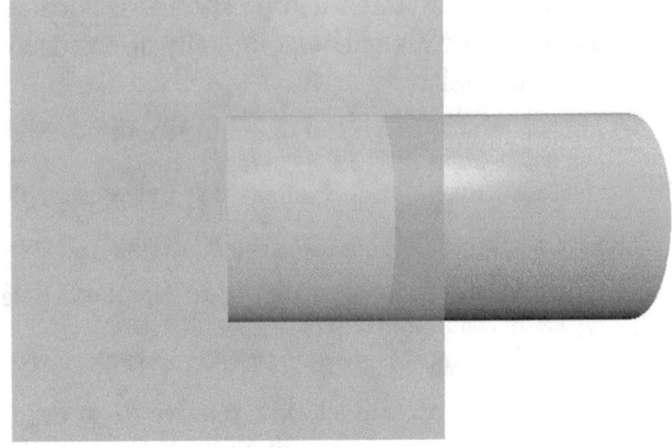

Fig. 9.6 Cube and cylinder with the same `pos` attribute.The cube is semitransparent

- **`length`** Length of `axis` and x component of `size`; default is `length=1`. Setting the length makes the magnitude of the axis equal to the length.
- **`size`** Instead of specifying `length` and `radius`, you can set `size=vector(length,height,width)`, which means that the cross section of the cylinder can be elliptical. Setting size makes the magnitude of axis be equal to the x component of the size (the length).

9.3.4 Cone

The cone object can have a circular or elliptical cross section and tapers to a point. The following commands create a cone with the center of its elliptical base at the origin, and its height parallel to the $(1, 0.2, -0.7)$ vector. The cone, shown in Fig. 9.7, has height 6, major axis 4 and minor axis 1. The second command actually creates a supplementary distant light source, independently of the object represented in the figure.

```
MyCone=cone(pos=vector(0,0,0),size=vector(6,4,1),color=vector(0.4,0.6,0.8),\
  axis=vector(1,0.2,-0.7))
distant_light(direction=vector(4,4,0),color=vector(0.5,0.5,0.5))
```

A circular cone can be created either by giving equal values to the last two components of the `size` vector, or by specifying `radius` and `length` rather than `size`.

9.3.5 Pyramid

The pyramid object has a rectangular cross section and tapers to a point. The following command creates a pyramid with its basis centered at the origin, The sides of the base are equal, 4 and 4, and the height is 3. The height is parallel to the $(-1, 0.2, 0.2)$ vector. The pyramid is shown in Fig. 9.8.

Fig. 9.7 Cone

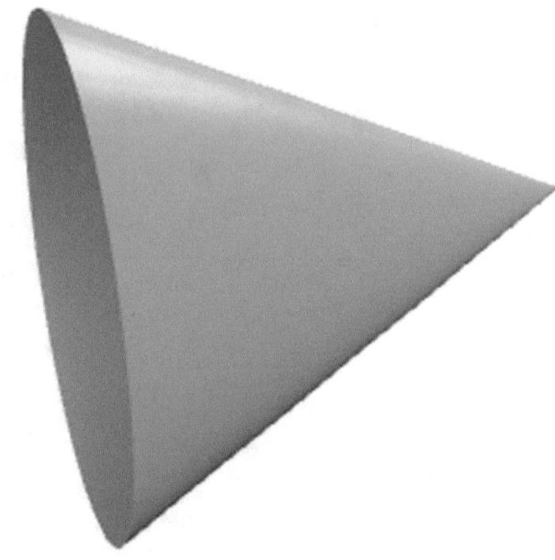

Fig. 9.8 Pyramid

```
MyPyramid=pyramid(pos=vector(0,0,0),size=vector(4,4,3),\
  color=vector(1,0.6,0.0),axis=vector(-1,0.2,0.7))
```

Instead of specifying the `size` vector, it is possible to specify `length`, `height` and `width`.

9.3.6 Ellipsoid

The ellipsoid object has the same attributes as the box object and it can be thought of as inscribed in the box specified by the same dimensions. The following command creates an ellipsoid with axes 6, 4 and 3, respectively, and with the firs axis parallel to the $(-1, 0.2, 0.7)$, as shown in Fig. 9.9.

Fig. 9.9 Ellipsoid

Fig. 9.10 Helix

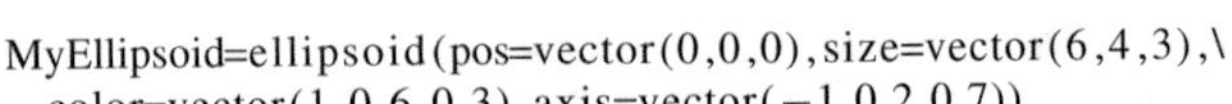

```
MyEllipsoid=ellipsoid(pos=vector(0,0,0),size=vector(6,4,3),\
  color=vector(1,0.6,0.3),axis=vector(-1,0.2,0.7))
```

The three axes can be specified separately as `length`, `height` and `width`.

9.3.7 *Helix*

The helix object has the following attributes and default values: `pos` (default: `vector(0,0,0)`), `axis` (default: `vector(1,0,0)`), `length` (default: 1), `radius` (default: 1), `coils` (default: 5), `thickness` (default: `radius`/20), `color` (default: vector(1,1,1)). The following command creates a helix with a "base" centered at the origin, length 6, an elliptical cross section of axes 4 and 3, respectively, and axis parallel to the vector (0, 1, 0.3). See Fig. 9.10

Fig. 9.11 Ring

```
MyHelix=helix(pos=vector(0,0,0),size=vector(6,4,3),color=vector(1,0.3,0.3),\
  axis=vector(0,1,0.3),coils=8)
```

9.3.8 Ring

The ring object is by default circular, with a specified radius and thickness and with its center given by the `pos` attribute:

```
MyRing=ring(pos=vector(0,0,0),size=vector(0.3,4,3),color=vector(1,0.3,0.3),\
  axis=vector(1,0,0.4))
```

Instead of specifying `size`, you can specify `radius` and `thickness`. Attribute `thickness` is the radius of the cross section of the ring, `radius` is the radius of the central part of the ring. Consequently the outer radius of the ring equals `radius+thickness`, and the inner radius is `radius-thickness` (Fig. 9.11).

9.3.9 Curve

The curve object is a polyline specified by the sequence of its vertices. If each vertex is sufficiently close to its two neighboring vertices you get the appearance of a smooth curve. You can start your curve by first specifying the coordinates of one of its ends, then appending the subsequent vertices one by one, as in Listing 9.4. Attribute `radius` is the radius of the cross section of the curve. Listing 9.4 first specifies the start point, color, and radius of the curve, then, the `while` loop adds the successive vertices. The result is shown in Fig. 9.12.

Listing 9.4 curve.py

```
Curve!=curve(vector(0,0,0),color=vector(0.3,0.3,1),radius=2)
r=0
while r<=150:
    x=r
    y=r*np.cos(0.2*r)
    z=r*np.sin(0.2*r)
    p=vector(x,y,z)
    Curve1.append(p)
    r+=1
```

It is also possible to specify the color and radius of the individual vertices of the polyline. In this case the color of each line segment will shade from the color of its initial vertex to the color of its final vertex, and the radius will change linearly from the initial to the final value. An example is shown in Fig. 9.13, generated by Listing 9.5.

Listing 9.5 CurveColor.py

```
1  line=curve(vector(0,0,0),color=vector(1,0,0),radius=4)
2  r=0
3  while r<=150:
4      x=r
5      y=r*np.cos(0.2*r)
6      z=r*np.sin(0.2*r)
7      p=vector(x,y,z)
8      if r<=50:
9          rr=1
10         gg=r/50
11         bb=0
12     elif r<=100:
13         rr=1-(r-50)/50
14         gg=1.0
15         bb=(r-50)/50
16     else:
17         rr=0
18         gg=1.0-(r-100)/50
19         bb=1
20     if r<=75:
21         rad=4+r*(8/75)
22     else:
23         rad=12-(r-75)*(8/75)
24     pd=dict(pos=p,color=vector(rr,gg,bb),radius=rad)
25     line.append(pd)
26     r+=1
```

Lines 8–19 evaluate the colors of the single vertices, while Lines 20–23 evaluate the radii. Line 24 create a *dictionary* (see Sect. 1.14) associating the values of position, color and radius of each vertex to the `keys` `pos`, `color` and `radius`. The dictionary is then appended to the curve at Line 25.

Fig. 9.12 Curve 1

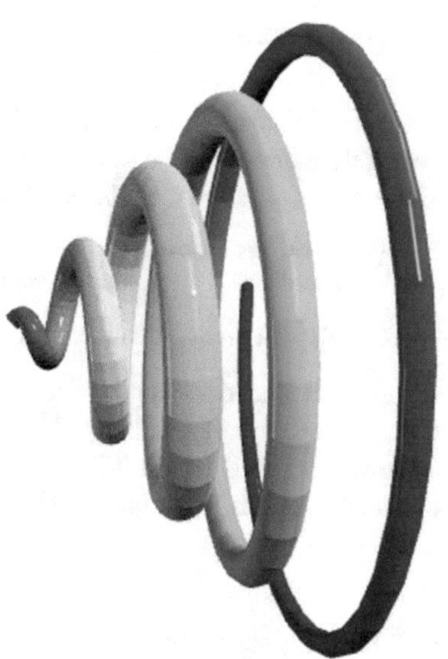

Fig. 9.13 Curve 2

If you dynamically append vertices to a curve in an animation, you can specify how many vertices should be retained. The following statement specifies that only the most recent 150 vertices should be retained as you add vertices to the curve:

```
line=curve(vector(-0,0,0),color=vector(1,0,0),radius=4,retain=150)
```

Fig. 9.14 Arrow

9.3.10 *Arrow*

The `arrow` object has a straight box-shaped shaft with an arrowhead (shaped as a pyramid) at one end. The following statement will display an arrow pointing parallel to the y axis, as shown in Fig. 9.14:

```
pointer=arrow(pos=vector(0,-3,0),length=6,axis=vector(0,1,0),\
  color=vector(0.9,0.7,0.3),shaftwidth=0.1)
```

The relevant arrow attributes are

- **`length`** The total length of the arrow, default is 1.
- **`shaftwidth`** By default 0.1*(length of arrow).
- **`headwidth`** By default 2*shaftwidth.
- **`headlength`** By default 3*shaftwidth.

9.3.11 *Compound*

The `compound` object lets you group objects together and manage them as if they were a single object. Moreover, displaying or moving a complicated compound object is faster than displaying or moving the individual objects one at a time.

The following code creates a hammer, with its handle parallel to the x axis and head parallel to the y axis, as a compound of three objects, namely a `cylinder`, a `box` and a `pyramid`. The result is shown in Fig. 9.15

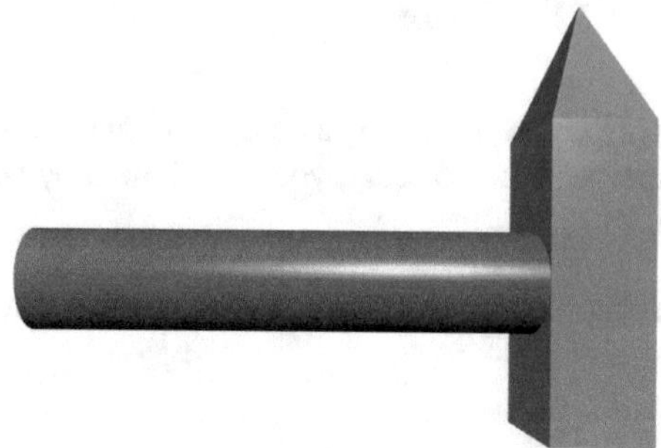

Fig. 9.15 Horizontal hammer

Fig. 9.16 Rotated hammer

```
handle=cylinder(pos=vector(-1,0,0),size=vector(2,0.4,0.4),\
  color=vector(0.72,0.42,0))
head=box(pos=vector(1.2,0,0),size=vector(0.4,1.2,0.4),\
  color=color.gray(0.6))
point=pyramid(pos=vector(1.2,0.6,0),size=vector(0.5,0.4,0.4),\
  axis=vector(0,1,0),color=color.gray(0.6))
hammer=compound([handle,head,point])
```

Once you have created a compound, you can manage it as if it were a single object. For instance, the following single command rotates the whole hammer, as shown in Fig. 9.16.

```
hammer.axis=vector(0.5,0.5,0)
```

9.3.12 Clone

Any object, including a compond object, can be cloned. For instance, once you have created your hammer, you can clone it by typing (Fig. 9.17)

```
ClonedHammer=hammer.clone(pos=vector(2,0,0))
```

After making the clone, changes to the clone do not affect the original object.

9.3.13 Points

The `points` object is essentially the same as the `curve` object except that it displays individual points at the specified positions instead of connecting them with lines. The `points` object does not have `origin`, `size`, or `axis` attributes, nor does it have a `rotate()` method.

Figure 9.18 is generated by Script 9.6, obtained from Script 9.5 simply by replacing the word `curve` by the word `points()` in Line 1, and by letting the point radius increase till the end of the loop. Note that points of the same radius appear smaller if they are farther from the camera, larger if they are closer to the camera.

Listing 9.6 points.py

```
1  line=points(vector(0,0,0),color=vector(1,0,0),radius=4)
2  r=0
3  while r<=150:
4      x=r
5      y=r*np.cos(0.2*r)
6      z=r*np.sin(0.2*r)
7      p=vector(x,y,z)
8      if r<=50:
9          rr=1
10         gg=r/50
11         bb=0
12     elif r<=100:
13         rr=1-(r-50)/50
14         gg=1.0
15         bb=(r-50)/50
16     else:
17         rr=0
18         gg=1.0-(r-100)/50
19         bb=1
20     rad=4+r*(8/75)
21     pd=dict(pos=p,color=vector(rr,gg,bb),radius=rad)
22     line.append(pd)
23     r+=1
```

Fig. 9.17 Cloned hammer

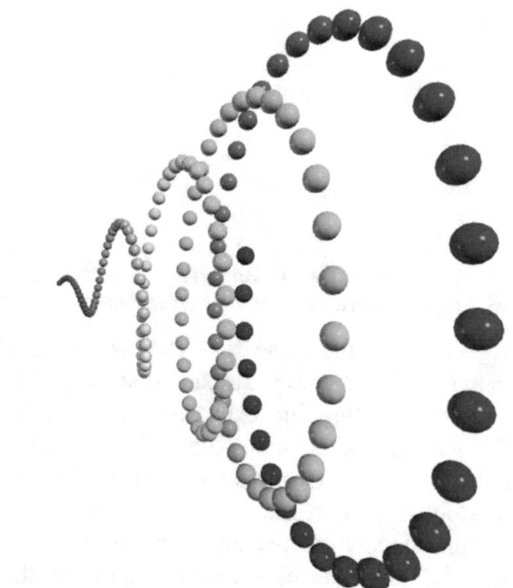
Fig. 9.18 Points

9.3.14 Extrusion

In real life, extrusion is a process used to create objects of a fixed cross-sectional profile. A material is pushed through a die of the desired cross-section. Extrusion applications range from metal pipes to good old Italian pasta. In VPython the `extrusion` object extrudes a 2D shape (the cross section) along a path. This is its minimal syntax:

```
MyExtr = extrusion(path=..., shape=...)
```

The `shape` attribute is a list of $[x, y]$ pairs describing the perimeter of the 2D cross section, while the `path` attribute is a list of points defining the path along which the cross section is extruded. This is a simple example:

Fig. 9.19 Extrusion

```
1  base=[[−1,1],[−1,−1],[0,−1.5],[1,−1],[1,1],[0,1.5],[−1,1]]
2  pth=[vector(0,0,0),vector(2,−1,4),vector(3,−4,5),vector(4,−5,10)]
3  MyExtr=extrusion(path=pth,shape=base,color=vector(0.4,1,0))
```

Line 1 creates the shape to be extruded, named `base`. Line 2 creates the extrusion path `pt`. The resulting extrusion is shown in Fig. 9.19.

After creating an extrusion named `MyExtr`, `MyExtr.size` represents the size of the bounding box of the extrusion. The x, y, and z components of `MyExtr.size` can also be referred to as `MyExtr.length`, `MyExtr.height`, and `MyExtr.width`.

Immediately after creation, the attribute `MyExtr.pos` comprises the coordinates of the center of the bounding box of the extruded object. Later changes to `MyExtr.pos` move the center of the extrusion to a new location. If you specify `pos` when creating the extrusion, the center of the extrusion will be moved to that location, just as if you changed `MyExtr.pos` after creating the extrusion.

Changing the extrusion `axis` rotates the combined object. As with some other objects such as `sphere` or compound, changing the axis does not affect the size.

VPython provides a *shapes and paths library* containing geometric figures to be used as contours of the cross-sections and as extrusion paths, see Sect. G.2 of the Appendix. Originally a shape is created in the xy plane. In using the paths library, it is helpful to think of the path as being first created in the xy plane. The path is then tipped back onto the xz plane, with the top of the path pointing in the $-z$ direction (into the computer monitor).

You can create a 2D shape with holes in it, in which case the extrusion has hollow sections. The code below creates an extrusion of an hexagon with a central circular hole.

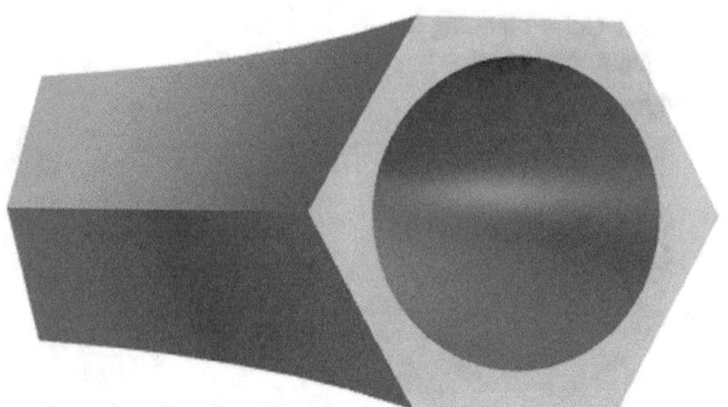

Fig. 9.20 Extrusion with hole

```
1 base=shapes.hexagon(length=1)
2 hole=shapes.circle(radius=0.7)
3 pth=paths.arc(radius=3, angle1=0, angle2=pi/3)
4 MyExtr=extrusion(shape=[base,hole],path=pth,color=vector(1.0,0.8,0.3))
```

Line 1 defines the extrusion cross section, Line 2 defines the hole, and Line 3 the extrusion path. The cross section and its hole are combined in Line 4. The result is shown in Fig. 9.20.

9.3.15 *Triangle/quad*

The object `triangle` is obviously a *triangle*, while the object `quad` is a *quadrilateral.* A `triangle` is defined by three `vertex` objects (its three vertices), and a `quad` is defined by four. The object `triangle` is very powerful because any surface can be approximated by a mesh of triangles of appropriate sizes, so you can make (actually, approximate) any shape you like. Further, by changing the attributes of the `vertex` objects, you can change the shape of a surface dynamically during an animation. A `vertex` object can be shared by several adjacent triangles, and changing the position of that vertex simultaneously affects all of the triangles sharing it.

You can create a `triangle` by first creating three `vertex` objects, then using them to form a triangle:

```
a=vertex(pos=vec(0,0,0))
b=vertex(pos=vec(1,0,0))
c=vertex(pos=vec(1,1,0))
T=triangle(v0=a,v1=b,v2=c)
```

Another option for the last command is to give a list of `vertex` objects:

```
T=triangle(vs=[a,b,c])
```

Often it is convenient to divide a surface into quadrilaterals instead of triangles, which is done with `quad` objects. The following code is equivalent to creating two triangles with vertices (a, b, c) and (a, c, d), respectively

Fig. 9.21 Color blending in the triangle created by Script 9.7

```
a=vertex(pos=vec(0,0,0))
b=vertex(pos=vec(1,0,0))
c=vertex(pos=vec(1,1,0))
d=vertex(pos=vec(0,1,0))
Q=quad(v0=a,v1=b,v2=c,v3=d)
```

You could also write this:

```
Q=quad(vs=[a,b,c,d])
```

Each vertex of a `triangle` or `quad` has its own color (default is white). If all vertices have the same color the surface of the `triangle` (`quad`) is uniformly colored. Otherwise each point of the surface has its own color, obtained by blending the colors of the vertices. This is shown in Fig. 9.21, generated by Script 9.7, where the three vertices of the equilateral triangle are given the three basic RGB colors: Line 14 assigns the red color to the upper vertex, Line 15 assigns the green color to the lower left vertex and Line 16 the blue color to the lower right vertex. As a result, each point of the triangle surface has its own color, with the three RGB color components blended according to how far the point is from each vertex: the closer the point is to one vertex, the stronger the component of the corresponding basic color.

Listing 9.7 vColorBlend.py

```
1 #!/usr/bin/env python3
2 #
3 from vpython import *
4 #
5 scene.width=600
6 scene.height=600
7 scene.background=vector(1,1,1)
8 distant_light(direction=vec(1,0,1),color=vec(1,1,1))
9 distant_light(direction=vec(-1,0,1),color=vec(1,1,1))
```

```
#
alpha=pi/6
h=cos(alpha)/2
xx=sin(alpha)
a=vertex(pos=vec(0,h,0),color=vec(1,0,0))
b=vertex(pos=vec(-xx,-h,0),color=vec(0,1,0))
c=vertex(pos=vec(xx,-h,0),color=vec(0,0,1))
T=triangle(v0=a,v1=b,v2=c)
```

An important attribute of a `vertex` is its `normal`. Actually, a `normal` is associated to each point of the `triangle` (or `quad`) surface, interpolated from the normals of the vertices. The normal of a surface point determines how light is reflected at that point, according to the laws of geometrical optics. The color of the reflected light depends on the color of the light source and on the color of the reflecting surface. In the simplest case, the normal at each vertex is perpendicular to the surface of the triangle.

Where two triangles meet, you will see a sharp break if the two triangles have different normals. If you want to smooth the transition, you should average the neighboring normals $(n_1 + n_2)/2$ and apply them to both triangles; this will smooth out the joint.

If you don't explicitly specify a normal for a vertex, the vertex normal is given the default value `vec(0,0,1)`, pointing toward you out of the screen. If the vertex is to be used in a triangle that does not lie in the xy plane, you need to specify the normal yourself. If the three vertices are `a`, `b`, and `c` (in counter-clockwise order), and you want normals perpendicular to the triangle surface, you can use the command

```
norm=(b.pos-a.pos).cross(c.pos-b.pos).norm()
```

since the vector cross product is perpendicular to the plane defined by the two vectors `(b.pos-a.pos)` and vector `(c.pos-b.pos)`, corresponding to the first two edges of the triangle as you go around it counterclockwise starting from vertex `a`.

As an example, Listing 9.8 uses triangles to create the octagonal pyramid shown in Fig. 9.22. The base of the pyramid lies on the $y = -1$ plane, with its center at `(0,-1,0)`, while the apex is located at `(0,1.5,0)`. As discussed in Sect. 9.4, you can rotate the "camera position" by moving the mouse pointer over the canvas while keeping the right mouse button pressed. In particular, if you rotate around the x axis, you can see the base (bottom) of the pyramid.

Listing 9.8 OctaPyram.py

```
#!/usr/bin/env python3
#
from vpython import *
#
scene.title='Octagonal Pyramid'
scene.width=1000
scene.height=600
scene.background=vector(1,1,1)
distant_light(direction=vec(1,2,1),color=vec(0.8,0.8,0.8))
distant_light(direction=vec(1,-2,1),color=vec(0.8,0.8,0.8))
#
```

```
col=vector(0.7,0.4,0)
pz=[]
px=[]
tria=[]
alpha=pi/4
yy=-1
```

Lists `pz` and `px` will contain the z and x coordinates of the vertices of the octagonal base, respectively, list `tria` will contain the triangles building the pyramid. Angle `alpha` is the central angle of the octagonal base, $\alpha = 2\pi/8 = \pi/4$ (45°), as shown in Fig. 9.23. Quantity `yy` is the y coordinate of the base.

Fig. 9.22 Octagonal pyramid built with triangles

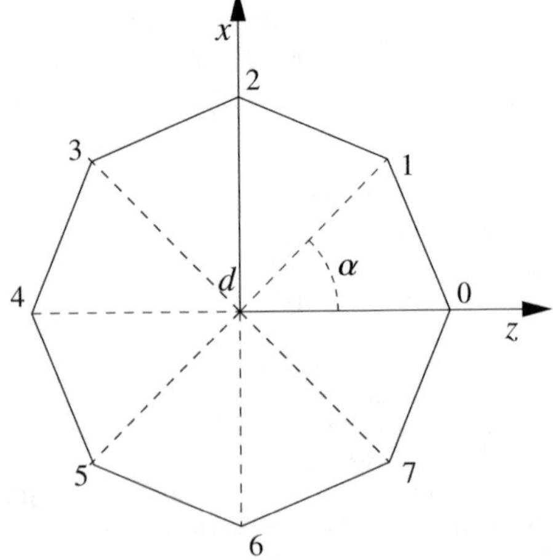

Fig. 9.23 Base of the octagonal pyramid on the $y = -1$ plane

```
18 for i in range(8):
19     beta=i*alpha
20     zz=cos(beta)
21     xx=sin(beta)
22     pz.append(zz)
23     px.append(xx)
24 c=vertex(pos=vec(0,1.5,0),color=col)
25 d=vertex(pos=vec(0,-1,0),color=col)
26 d.normal=vec(0,-1,0)
```

The loop 18–23 evaluates the z and x coordinates of the vertices of the octagonal base of the pyramid, as shown in Fig. 9.23. Line 24 defines the vertex `c`, the apex of the pyramid, shared by all triangles of the lateral surface, see Fig. 9.24. Line 25 defines the vertex `d`, shared by all triangles forming the base of the pyramid, as shown in Figs. 9.23 and 9.24. Line 26 defines the normal of vertex `d` as directed downwards, pointing out of the base of the pyramid.

```
27 for i in range(8):
28     j=(i+1)%8
29     al=vertex(pos=vec(px[i],yy,pz[i]),color=col)
30     ab=vertex(pos=vec(px[i],yy,pz[i]),color=col)
31     bl=vertex(pos=vec(px[j],yy,pz[j]),color=col)
32     bb=vertex(pos=vec(px[j],yy,pz[j]),color=col)
33     ab.normal=vec(0,-1,0)
34     bb.normal=vec(0,-1,0)
35     trb=triangle(v0=ab,v1=bb,v2=d)
36     lnorm=((bl.pos-al.pos).cross(c.pos-bl.pos)).norm()
37     al.normal=lnorm
38     bl.normal=lnorm
39     c.normal=lnorm
40     trl=triangle(v0=al,v1=bl,v2=c)
41     tria.append(trl)
42     tria.append(trb)
```

Loop 27–42 creates the triangles that form the pyramid. Each edge of the octagonal base is shared by two triangles, as shown in Fig. 9.24 for the edge ab. One triangle belongs to the lateral surface of the pyramid, with third vertex at the apex, `c`. The

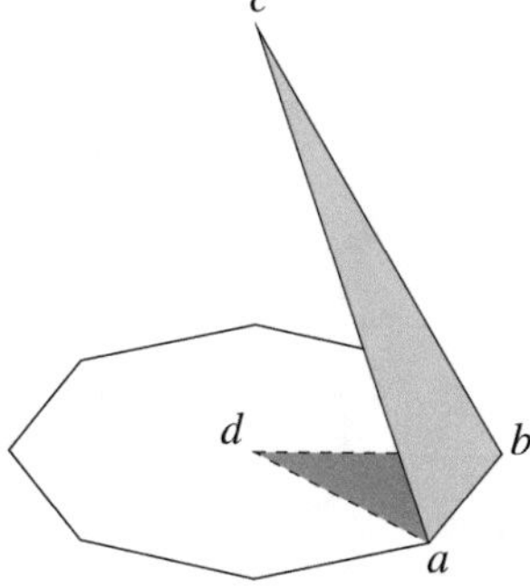

Fig. 9.24 Triangles building the octagonal pyramid

other triangle belongs to the base, with third vertex at the center of the octagonal base, d. In Lines 29–32 the vertices a1 and ab share the same position, but are given different normals at Lines 33 and 37, because a1 will serve as vertex of the "lateral" triangle, and ab as vertex of the "base" triangle. The same holds for vertices b1 and bb. Line 35 builds the base triangle, Line 45 the lateral triangle. List tria comprises al triangles building the pyramid.

If you add the following command

```
43 Pyram8=compound(tria)
```

you turn the pyramid into a single compound object, named Pyram8, that you can move and resize and rotate very rapidly, as if it were a VPython built-in object, such as a sphere.

9.3.16 Label

With the label object you can display text in a box. A label always faces forward, even if you rotate the camera (see the text object for 3D text). Here is a simple example:

```
MyEllipsoid=ellipsoid(pos=vector(0,0,0),size=vector(4,3,3),\
  color=vector(1,0.8,0.3))
MyLab=label(pos=vector(0,2,0),text='This is an ellipsoid',height=40)
```

The result is shown in Fig. 9.25. A unique feature of the label object is that several attributes are given in terms of *screen pixels* instead of the usual "world-space" coordinates. For example, the height of the text is given in pixels, with the result that the text remains readable even when the ellipsoid object is moved far away, its image becoming consequently smaller. Other pixel-oriented attributes include xoffset, yoffset, and border. Here are the label attributes

- **pos** The coordinates of the label location.

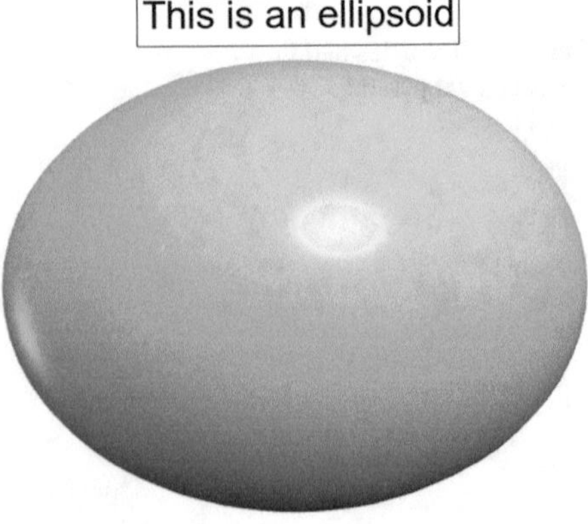

Fig. 9.25 Label

- **pixel_pos** If *True*, `pos` is the position in terms of pixels, with `vec(0,0,0)` in the lower left corner.
- **align** Specify `'left'`, `'right'`, or `'center'`. When you specify `'center'`, which is the default, the center of the topmost line of text is at the location given by `pos`. When you specify `'left'`, the start of the topmost line of text is at the location given by `pos`. When you specify `'right'`, the right end of the topmost line of text is at the location given by `pos`.
- **xoffset**, **yoffset** The x and y offsets of the line from `pos`, in pixels. Unless you specify a value for `align`, text above or below `pos` is centered, text to the left of `pos` is right-adjusted, and text to the right of `pos` is left-adjusted.
- **text** The text to be displayed, such as `'This is an ellipsoid'`. You can insert numbers or numerical expressions; the whole will be converted to a string.You can include the HTML styles for italic (<i> or <em>), bold (<b> or <strong>), superscript (<sup>), or subscript (<sub>). For example, the string

  ```
  'The␣<b>mass␣<i>M</i></b><sub>sys</sub>␣=␣10<sup>3</sup>␣kg.'
  ```

  displays as

  'The **mass** M_{sys} = 103 kg.'

  Multiple lines can be displayed by inserting line breaks (\ n), as in 'Three\ n lines\ n of text', alternatively you can use <br> or <br/ >, according to HTML style. You can also create multiple lines by using triple quotes:

  ```
  L = label()
  L.text = """The quick brown fox
  jumps over the lazy dog."""
  ```

- **font** Name of the desired font; `'sans'` or `'serif"` or `'monospace'` (fixed-width), or a specific font name; the default is `"sans"`
- **height** Height of the font in pixels; default is 15 pixels.
- **color** Color of the text; default is scene.foreground.
- **background** Color of the background of the box; default is scene.background.
- **opacity** Opacity of the background of the box, default 0.66 (0 transparent, 1 opaque, for objects behind the box).
- **border** Distance in pixels from the text to the surrounding box; default is 5 pixels
- **box** *True* if the box should be drawn (default), else *False*
- **line** *True* if a line from the `pos` to the box should be drawn (default), else *False*.
- **linecolor** Color of the `line` and `box`
- **linewidth** Thickness of the line drawn from the `pos` to the `box`, and the edges of the `box` (default is 1 pixel)
- **space** Radius in pixels of a sphere surrounding `pos`, into which the connecting line does not go
- **visible** Label is not displayed if *False*.

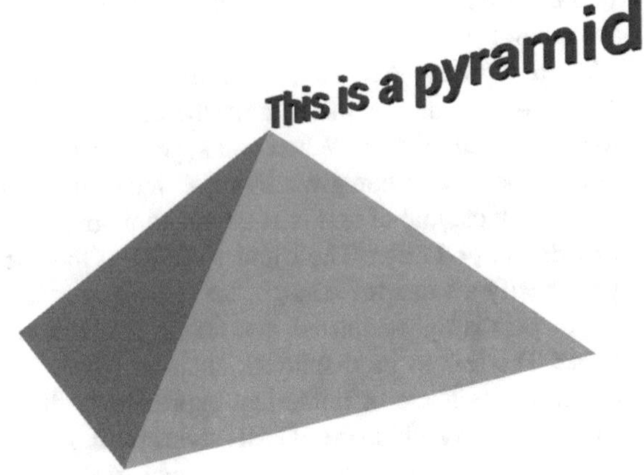

Fig. 9.26 3D Text

9.3.17 *Text*

With the `text` object you can display 3D text. The blue 3D text appearing in Fig. 9.26 was created by the following code

```
1 MyPyramid=pyramid(pos=vector(0,-1,0),size=vector(2,4,4),\
2    axis=vector(0,1,0),color=vector(1,0.6,0.3))
3 MyPyramid.rotate(angle=-45)
4 MyText=text(pos=vector(0,1,0),text='This is a pyramid',height=0.3,\
5    color=vector(0,0.4,1),axis=vector(1,0,1))
```

where Lines 1–2 create a pyramid with its base centered at $(0, -1, 0)$ and height parallel to the y axis. Line 3 rotates the pyramid about its axis by $-45°$, and Lines 4–5 create the 3D text parallel to the $(1, 0, 1)$ vector.

Currently this 3D text object does not handle HTML options, such as bold, italic, subscript, or superscript. The label object does however support these options.

9.4 Controlling the View

You can interpret what you see in the VPython canvas as images taken by a video camera located, by default, on the z axis of the canvas reference frame and pointing toward the origin, as shown in Fig. 9.27. By default the camera is located at $(0, 0, \sqrt{3})$. You can use the mouse to change the camera position. If you turn the mouse scroll wheel the camera moves back and forward along the z axis (line **a** of Fig. 9.27), correspondingly you see the object becoming larger or smaller. If you move the mouse cursor horizontally on the canvas while keeping the <Shift> key of the keyboard and the left mouse button simultaneously pressed, the camera moves along the x axis (line **b**). Analogously, if you move the mouse cursor vertically while pressing <Shift> and the left mouse button the camera moves along the y axis (line **c**). Finally, if you

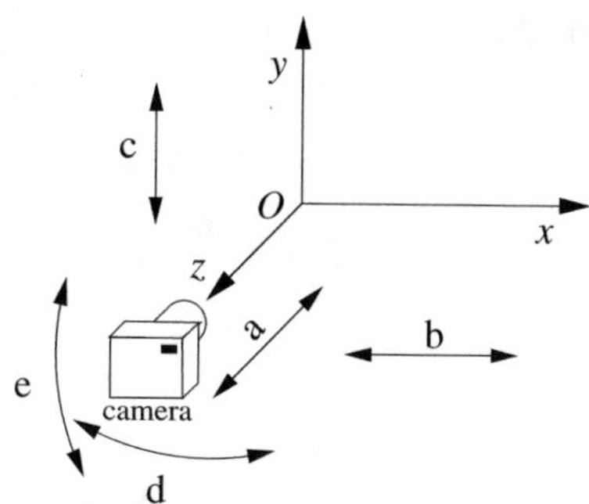

Fig. 9.27 Using the mouse to move the camera

keep the right mouse button pressed and move the mouse horizontally the camera rotates around the y axis, always pointing toward the origin (arc **d** of Fig. 9.27)). Analogously, if you move the mouse vertically the camera rotates around the x axis (arc **e**).

If you are not satisfied with the default position of the camera, you can place the camera where you want by typing the command

```
scene.camera.pos=vector(my_x,my_y,my_z)
```

where `my_x`, `my_y` and `my_z` are the Cartesian coordinates of the selected camera location. This will leave the camera still pointing in the $-z$ direction. You can change the direction the camera is pointing with the command

```
scene.camera.axis=vector(my_x,my_y,my_z)
```

where `my_x`, `m_y` and `my_z` are the components of the required direction.

You can change the canvas size with the commands

```
scene.width=n_w
scene.height=n_h
```

where `n_w` and `n_h` are the numbers of pixels required for the canvas width and for the canvas height, respectively. As previously mentioned, if you are not satisfied with the black color of the canvas background you can change it, for instance, with the command

```
scene.background=color.white
```

where `color.white` can be replaced by any color listed in Table G.1 of the Appendix, or by any color you can create withe the RGB additive color model described in Sect. G.1 of the Appendix. Finally, you can give the canvas a title by typing

```
scene.title='MyTitle'
```

as shown in Fig. 9.28.

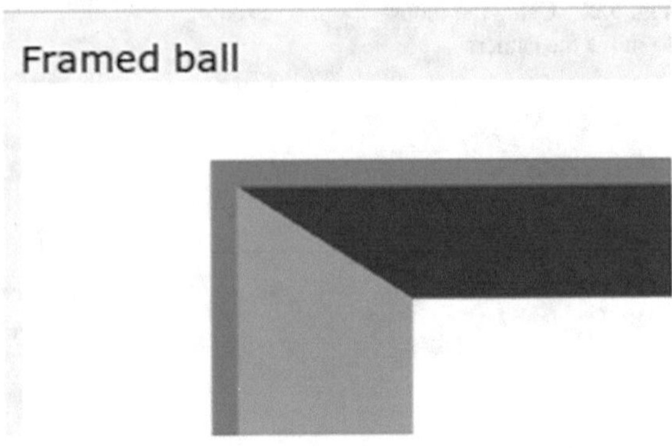

Fig. 9.28 Canvas title

As mentioned in Sect. 9.3.1, by default the scene is illuminated by two distant light sources, which you can switch off by the command

```
scene.lights=[]
```

You can add as many *distant* or *local* light sources (*distant* meaning that the light rays propagate parallel to one another, *local* that the light rays originate from a point light source) as you wish by commands of the form

```
1 FarLight1=distant_light(direction=vec(0.22,0.44,0.88),color=color.gray(0.8))
2 lamp1=local_light(pos=vec(x,y,z),color=color.yellow)
```

where in Line 1 the arguments of `vec()` are the vector components of the direction the light is *coming from* (so, the light propagates in the opposite direction), while in Line 2 `pos` is the location of the point light source. The light color is specified according to Sect. G.1 of the Appendix.

You are invited to do your own experiments with the commands described in this section, since practice is the best possible teacher.

9.5 VPython Animation

9.5.1 Basics

As discussed in Sect. 7.1 for Tkinter, also VPython animation is obtained by displaying successive still images (frames) at a given frame rate. Thus we need a loop that slightly changes the position and/or orientation of one or more objects on the canvas at every iteration. As a first example, Listing 9.9 is a more refined VPython equivalent of Listing 7.1, showing a ball bouncing on the walls of a room in the presence of gravity (and in the absence of any kind of friction!)

Line 2 imports everything from VPython, Line 3 writes a small title above the upper left corner of the canvas, the detail is shown in Fig. 9.28, while Fig. 9.29 shows the full canvas. Line 4 sets the canvas background to white, Lines 6 and 7 specify the canvas width and height in pixels.

Listing 9.9 VFramedBall.py

```
1  #!/usr/bin/env python3
2  from vpython import *
3  # ............................................... create canvas
4  scene.title='Framed ball'
5  scene.background=vector(1,1,1)
6  scene.width=1000
7  scene.height=600
8  # .................................................... variables
9  rw=8
10 rh=5
11 rdepth=1
12 thick=0.1
13 rad=0.2
14 vx=3
15 vy=8
16 dt=0.01
17 g=vector(0,-9.8,0)
```

Lines 9–11 define the width, height and depth, respectively, of the room inside which the ball will bounce. Line 12 defines the thickness of the room walls (the room walls have a thickness, see Fig. 9.29, thus they are cuboids, called “boxes” by VPython). All values are assumed to be in meters, even if this is irrelevant to our purposes. Line 13 specifies the radius of the bouncing ball, Lines 14 and 15 its initial x and y velocity components. Variable `dt` is the “real world” time interval between two successive animation frames, and `g` is the gravity acceleration.

```
18 # ..................................................... add lamp
19 lamp=local_light(pos=vector(1.2*rw,0,4*rdepth),color=color.white)
20 # ................................................... boundaries
21 LeftWall=box(pos=vector(-rw/2-thick/2,0,0),length=thick,\
22    height=rh+2*thick,width=rdepth,color=vector(0.5,0.5,1))
23 RightWall=box(pos=vector(rw/2+thick/2,0,0),length=thick,\
24    height=rh+2*thick,width=rdepth,color=vector(0.5,0.5,1))
25 floor=box(pos=vector(0,-rh/2-thick/2,0),length=rw,\
26    height=thick,width=rdepth,color=vector(0.5,0.5,1))
27 ceiling=box(pos=vector(0,rh/2+thick/2,0),length=rw,\
28    height=thick,width=rdepth,color=vector(0.5,0.5,1))
29 # ......................................................... ball
30 ball=sphere(pos=vector(-rw/2+rad,-rh/2+rad,0),radius=rad,\
31    make_trail=True,retain=50,color=color.red,\
32       velocity=vector(vx,vy,0))
```

Line 19 adds an auxiliary lamp. You can experiment what happens if you change its position and/or color. Lines 21–28 create the left and right walls, floor and ceiling of the room. In interpreting the commands, remember that, for each “box” (cuboid), length is parallel to the x direction, height to the y direction and width to the z direction.

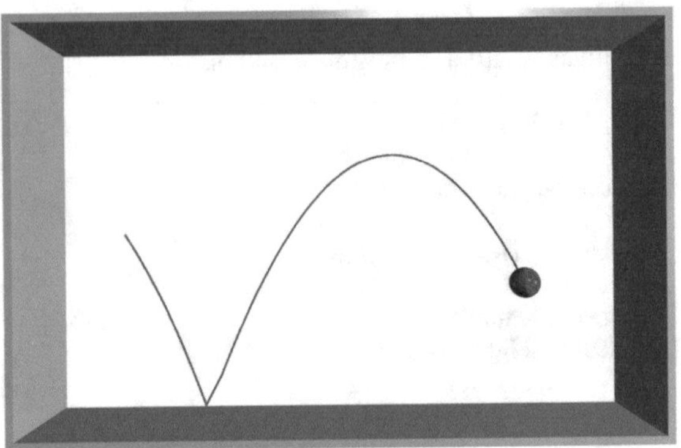

Fig. 9.29 Ball bouncing inside a box

Lines 30–32 create the ball in its initial position. The ball is initially located at the lower left corner of the room. Attribute `make_trail=True` means that the ball will leave a trail when moving. Attribute `retain=50` means that the trail will retain the positions of the last 50 frames. An additional, optional attribute is `trail_type`. If you set `trail_type='points'` the trail of the ball consists of individual points rather than a continuous curve. Default is `trail_type='curve'`. Quantity `ball.velocity` is the initial velocity of the ball.

```
33 # ...................................................... main loop
34 while True:
35   rate(50)
36   ball.pos=ball.pos+ball.velocity*dt+0.5*g*dt**2
37   ball.velocity=ball.velocity+g*dt
38   # .................... is the ball bouncing on the canvas borders?
39   if ball.pos.x>=(rw/2-rad):
40     ball.velocity.x=-abs(ball.velocity.x)
41   elif ball.pos.x<=-(rw/2-rad):
42     ball.velocity.x=abs(ball.velocity.x)
43   elif ball.pos.y>=rh/2-rad:
44     ball.velocity.y=-abs(ball.velocity.y)
45   elif ball.pos.y<=-(rh/2-rad):
46     ball.velocity.y=abs(ball.velocity.y)
```

Lines 34–46 constitute the animation loop. The command `rate(50)` halts computations long enough to make sure that at least a time interval of 1.0/50.0 = 0.02 s has elapsed between two successive animation frames. If this much time has already elapsed, no halt is performed. If you place `rate(50)` inside a computational loop, the loop will execute at a maximum of 50 times per second, even if the computer can run faster than this. This makes animations look about the same on computers of different speeds, as long as the computers are capable of carrying out 50 computations per second. It is important to remark that, if you are running your programs on the `glowscript.org` web site, any infinite loop *must* contain a `rate()` command, otherwise your program will lock up the browser. The `rate()` command provides the pause that makes further updates to the page possible. Line 36 updates the ball position according to the law of motion with uniform acceleration, while Line 37 updates the ball velocity. Lines 39–46 invert the appropriate component of the ball velocity when the ball is hitting one of the box walls.

Motion can involve rotations, including revolutions of objects around their axes. As an example, Script 9.10 generates an arrow (see Sect. 9.3.10 and Fig. 9.14) revolving around its axis own and precessing around the y axis of the canvas.

Listing 9.10 VPrecession.py

```
1  #!/usr/bin/env python3
2  from vpython import *
3  scene.title='Precession'
4  scene.width=800
5  scene.height=600
6  scene.background=vector(1,1,1)
7  # .......................................... Initial conditions
8  s=2
9  theta=pi/6
10 phi=0.0
11 sy=s*cos(theta)
12 r=s*sin(theta)
13 spin=arrow(pos=vector(0,0,0),length=s,axis=vector(0,sy,r),\
14   color=vector(1,1,0))
15 dPhi=radians(0.5)
16 dPsi=radians(3)
17 # .................................................... Precession
18 while True:
19   rate(50)
20   spin.rotate(angle=dPsi)
21   sz=r*cos(phi)
22   sx=r*sin(phi)
23   spin.axis=vector(sx,sy,sz)
24   phi+=dPhi
```

Lines 8–14 define the initial conditions of our precession motion. Quantity `s` at Line 8 is the length of the arrow representing our precessing spin, in arbitrary units. Initially the spin lies in the yz plane forming an angle `theta` ($\vartheta = \pi/6$, or 30°) with the y axis. Angles `theta` and `phi`, initially $\varphi = 0$, are the usual spherical angular coordinates. The only difference is that here, for convenience, the polar axis is y, and φ is measured from the z axis in the zx plane, as shown in Fig. 9.30. The angle ψ is the revolution angle of the arrow around its own axis.

Thus, the constant component of **s** along y is `sy`, $s_y = s \cos\vartheta$, and the projection of **s** onto the xz plane is `r`, $r = s \sin\vartheta$. Lines 13–14 create a yellow arrow of length `s` representing our precessing spin in its initial position. Lines 15 and 16 specify `dPhi` and `dPsi`, the increments $\Delta\varphi$ of φ and $\Delta\psi$ of ψ, respectively, at each iteration of the animation loop. Function `radians()`, provided by the `vpython` package, converts angle values from degrees to radians.

Lines 18–24 constitute the actual animation loop. Line 20 rotates the spin by `dPsi` around its own axis. Lines 21 and 22 evaluate the spin components on the xz plane, `sx` and `sz`, the y component being constant. Line 23 moves the spin axis to its new position, and Line 24 increments the φ coordinate for the next iteration.

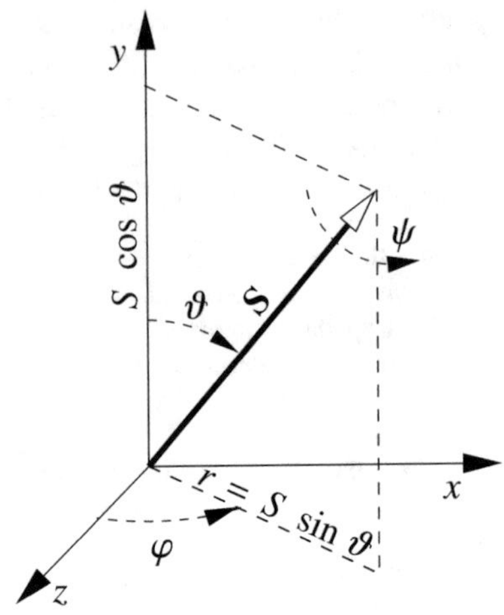

Fig. 9.30 Precession motion

9.5.2 *Interactive Control*

In Section 7.4 we learned how to interact with a Tkinter animation. Also VPython animations can be controlled by buttons and entries. Listing 9.11 adds interactive control to Listing 9.9. Only the changes from Listing 9.9 are commented.

Listing 9.11 VButtonBall.py

```
#!/usr/bin/env python3
from vpython import *
# .................................................. create canvas
scene.title='Buttons and Entries'
scene.background=vector(1,1,1)
scene.width=1000
scene.height=600
# ...................................................... variables
bw=8
bh=5
bdepth=1
thick=0.1
rad=0.2
vx=3
vy=8
dt=0.01
g=vector(0,-9.8,0)
# ........................................... interaction variables
RunIter=False
ButtonWidth=20
global MyText
```

The ball will be moving only if variable `RunIter` is *True*, otherwise it will be in standby. Variable `ButtonWidth` is the width, in characters, of the start/stop button. This is needed to keep the width of the button, bound to the function `StartStop()`

defined at Lines 34–40 below, constant. The lengths of the words *Stop* and *Restart* being different, the size of the button would change when its label changes from one word to the other, unless we force it to be constant. Variable `MyText` is a global variable used by functions `ReadEntry()` and `EraseText()`, defined below.

```
22 # ......................................................functions
23 def incvx():
24   ball.velocity.x=ball.velocity.x*1.2
25 def decvx():
26   ball.velocity.x=ball.velocity.x/1.2
27 def ReadEntry(s):
28   global MyText
29   MyText=text(pos=vector(0,1,0),text=s.text,height=0.3,\
30     color=vector(0.3,0.8,0))
31 def EraseText():
32   global MyText
33   MyText.visible=False
34 def StartStop():
35   global RunIter
36   RunIter=not RunIter
37   if RunIter:
38     StopButton.text='Stop'.center(ButtonWidth)
39   else:
40     StopButton.text='Restart'.center(ButtonWidth)
41 # ....................................... buttons and input window
42 AccelButton=button(bind=incvx,text='Faster!')
43 scene.append_to_caption('␣')
44 DecelButton=button(bind=decvx,text='Slower!')
45 scene.append_to_caption('\n')
46 StopButton=button(bind=StartStop,text='Start'.center(ButtonWidth))
47 scene.append_to_caption('\n\n')
48 LabelText=wtext(text='Write␣here:␣')
49 EntryWindow=winput(bind=ReadEntry,type='string',width=200)
50 scene.append_to_caption('␣')
51 EraseButton=button(bind=EraseText,text='Erase')
```

Functions `incvx()` and `decvx()` respectively increase and decrease the x component of the ball velocity, by multiplying or dividing it by the factor 1.2. Function `ReadEntry()` writes a text on the canvas (see Sect. 9.3.17) according to what you type in the input window. Function `EraseText()` erases the same text from the canvas by making the global variable `MyText` invisible. Function `StartStop()` switches the value of the variable `RunIter` from *False* to *True* and vice versa, and changes the label of the Start/Stop button accordingly.

Line 42 creates the acceleration button, bound to the function `incvx()`. When this button is pressed the horizontal component of the ball velocity is multiplied by 1.2. The button label is `Faster!`. By default, widgets, including buttons, are appended to `scene.caption_anchor`, below the canvas, as shown in Fig. 9.31. Line 43 appends a space to the caption area, in order to separate the acceleration button from the deceleration button, created at Line 44. Line 45 appends a "line feed", so that the next button will be placed in the next line below. Line 46 creates the `StopButton`, bound to function `StartStop()`.

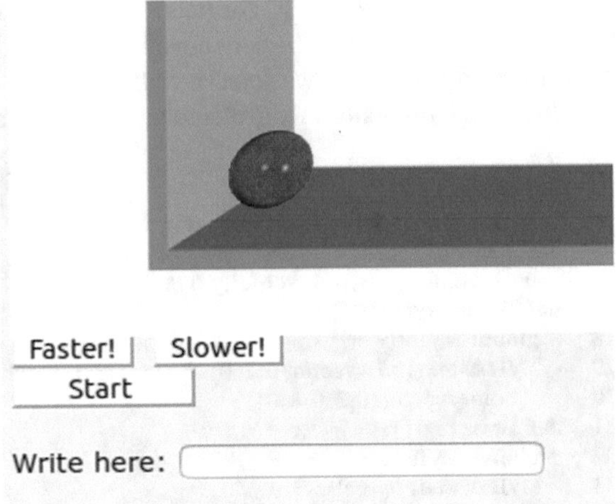

Fig. 9.31 The caption area below the canvas. By default, buttons and other widgets are appended here

Line 47 appends two line feeds, in order to leave one blank line between the control buttons and the entry widget. Line 48 writes the label `Write here:` before the entry window created at Line 49. If you type something in the entry window and press <Return>, what you wrote will be read and copied into variable `MyText`, which will be written on the canvas by function `ReadEntry()`. Line 51 creates the erase button, bound to function `EraseText()`. When you press the erase button, function `EraseText()` makes `MyText` invisible. Typing a new string in the entry window copies the new string into `MyText`, and makes it visible again.

```
52  # ........................................................ add lamp
53  lamp=local_light(pos=vector(1.2*bw,0,4*bdepth),color=color.white)
54  # .................................................... boundaries
55  LeftWall=box(pos=vector(-bw/2-thick/2,0,0),length=thick,\
56     height=bh+2*thick,width=bdepth,color=vector(0.5,0.5,1))
57  RightWall=box(pos=vector(bw/2+thick/2,0,0),length=thick,\
58     height=bh+2*thick,width=bdepth,color=vector(0.5,0.5,1))
59  floor=box(pos=vector(0,-bh/2-thick/2,0),length=bw,\
60     height=thick,width=bdepth,color=vector(0.5,0.5,1))
61  ceiling=box(pos=vector(0,bh/2+thick/2,0),length=bw,\
62     height=thick,width=bdepth,color=vector(0.5,0.5,1))
63  # ............................................................ ball
64  ball=sphere(pos=vector(-bw/2+rad,-bh/2+rad,0),radius=rad,\
65     make_trail=True,retain=50,color=color.red,\
66        velocity=vector(vx,vy,0))
67  # ..................................................... main loop
68  while True:
69     rate(50)
70     if RunIter:
71        ball.pos=ball.pos+ball.velocity*dt+0.5*g*dt**2
72        ball.velocity=ball.velocity+g*dt
```

Differently from Script 9.9, here the ball position and velocity are updated only if `RunIter` is *True*, otherwise the ball is in stand by.

```
    # .................. is the ball bouncing on the canvas borders?
    if ball.pos.x>=(bw/2-rad):
      ball.velocity.x=-abs(ball.velocity.x)
    elif ball.pos.x<=-(bw/2-rad):
      ball.velocity.x=abs(ball.velocity.x)
    elif ball.pos.y>=bh/2-rad:
      ball.velocity.y=-abs(ball.velocity.y)
    elif ball.pos.y<=-(bh/2-rad):
      ball.velocity.y=abs(ball.velocity.y)
```

Further available widgets, not used in the above example, are

- **`slider`** You drag the slider to enter a numeric value graphically.
- **`menu`** A pull-down menu of options
- **`checkbox`**
- **`radio`** Radio buttons are often used in a group of two or more, in which you arrange that only one is checked at any time.

each of which can be bound to a function of the program.

9.6 2D Graphics

VPython also offers the possibility to draw *graphs*. A *graph* is different from a canvas: a canvas is inherently 3D and contains 3D objects such as spheres and boxes, whereas a graph is inherently 2D and contains graphing objects for displaying curves, dots, and vertical or horizontal bars. Listing 9.12 gives a first example in interactive mode, the result is shown in Fig. 9.32.

Listing 9.12 Basic Plot

```
1 >>> from vpython import *
2 >>> fx=gcurve(color=color.red)
3 >>> for x in arange(0, 8.05, 0.1):
4 ...     fx.plot(x,5*cos(2*x)*exp(-0.2*x))
5 ...
6 >>> fx.plot(9,2.5)
```

Line 2 declares `fx` to be a connected curve (`gcurve`), plotted in red. If no color is specified, the default plotting color is black. Alternatives to a connected curve are disconnected dots, (`gdots`) vertical bars (`gvbars`), and horizontal bars (`ghbars`). Lines 3–4 constitute a loop where x goes from 0 to 8 with a step of 0.1. Each iteration of the loop evaluates the function $f(x) = 5\cos(2x)\,\mathrm{e}^{-0.2x}$ and adds a point of coordinates $[x, f(x)]$ to the curve. Line 5, being empty, ends the loop. If you stop here, skipping Line 6, you obtain Fig. 9.32 in the x range [0, 8] (thus, the figure stops at $x = 8$, and the last vertical strip from $x = 8$ to $x = 9$ is missing). If you now type Line 6, the point of coordinates (9, 2.5) is added to the plotted points, and

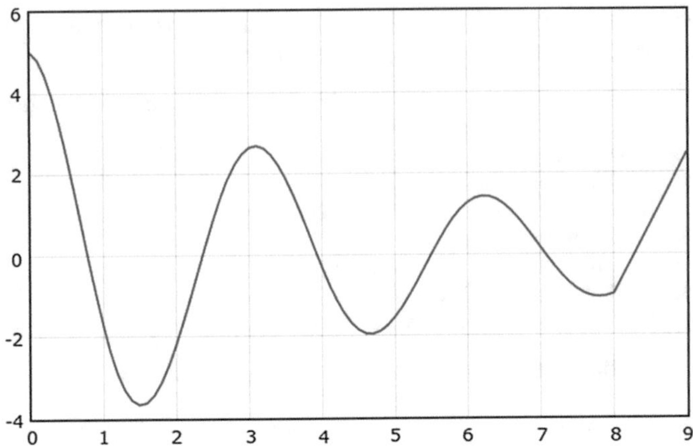

Fig. 9.32 Plot generated by Listing 9.12

the line segment from the last previous point, (8, −0.967), and (9, 2.5) is added. An equivalent way of typing Line 6 is `fx.plot([9,2.5)]`. You can add a further point

```
>>> fx.plot([new_x,new_y])
```

or multiple subsequent points by typing

```
>>> fx.plot([new_x1,new_y1],[new_x2,new_y2],[new_x3,new_y3], ...)
```

Once you have created a graphing object such as gcurve, its initial color will be used throughout when you add further points. You can change the color later, but this will change the color of all the elements already plotted.

You can obtain a vertical-bars plot like the one of Fig. 9.33 by running the Script 9.13

Listing 9.13 Vertical Bars

```
1 #!/usr/bin/env python3
2 from vpython import *
3 vb=gvbars(color=color.blue,delta=0.1)
4 for x in arange(-2,2.05,0.2):
5   vb.plot(x,x**2)
```

The option `delta=0.1` at Line 3 specifies the width of the vertical bars, default is 1. Listing 9.14 generates the horizontal-lines plot of Fig. 9.34

Listing 9.14 Horizontal Bars

```
1 #!/usr/bin/env python3
2 from vpython import *
3 hb=ghbars(color=color.blue,delta=0.05)
4 for x in arange(0,2.05,0.2):
5   hb.plot(x,sqrt(x))
```

here, at Line 3, `delta` specifies the height of the horizontal bars.

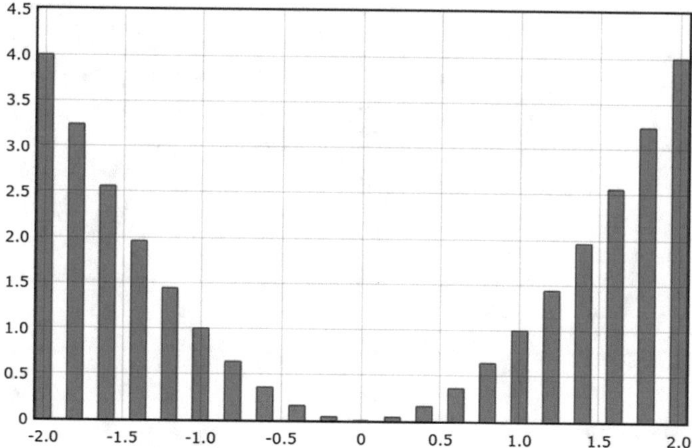

Fig. 9.33 Plot generated by Script 9.13

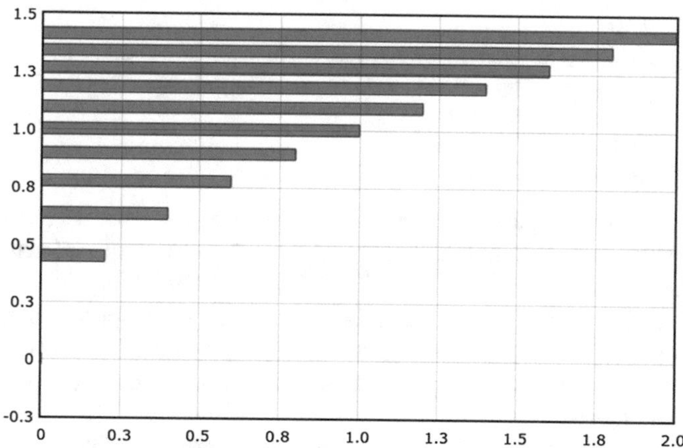

Fig. 9.34 Plot generated by Script 9.14

You can draw more than one graphing object in the same window, as shown in Fig. 9.35, generated by Listing 9.15, and displaying a damped cosine wave and an exponential. For each graphing object you can specify a label, that will appear as a text, in the same color of the object, at the upper right of the window, again as in Fig. 9.35. If you type the option `legend=False` the label will actually not be shown. Options like `label` must be specified when you create the graphing object. Once you have created a graphing object, its initial color will be used throughout. You can change the color later, but this will change the color of all the elements already plotted.

Listing 9.15 Damped Cosine

```
#!/usr/bin/env python3
from vpython import *
```

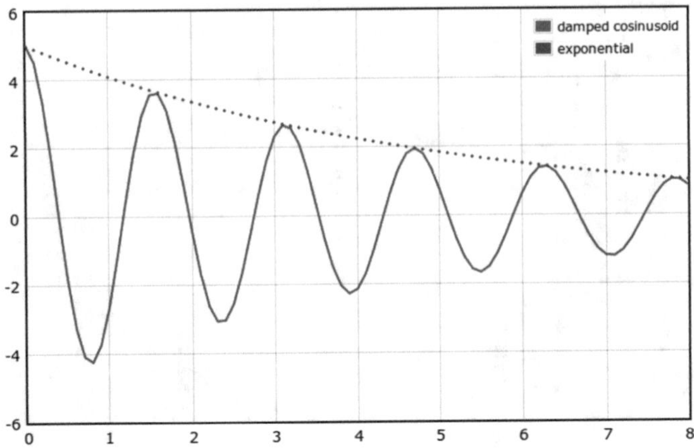

Fig. 9.35 Plot generated by Listing 9.12

```
3 fx=gcurve(color=color.red,label='damped␣cosinusoid')
4 gx=gdots(color=color.blue,label='exponential',radius=1)
5 for x in arange(0, 8.05, 0.1):
6   fx.plot(x,5*cos(4*x)*exp(-0.2*x))
7   gx.plot(x,5*exp(-0.2*x))
```

Line 2 defines `fx` as a graphing object of the type `gcurve`, that will represent the damped cosine wave. The option `label=damped_cosinusoid` generates the text at the upper right of Fig. 9.35. Line 3 defines `gx` as a graphing object of the type `gdots`, that will represent the exponential determining the amplitude of the cosine wave. The option `radius=1` specifies the radius, in px, of the dots forming the plot, default is `radius=3`. The loop 5–7 evaluates the coordinates of the points of the two graphing objects.

When creating a graph, you can specify logarithmic plots by specifying `logx=True` and/or `logy=True`. All values must be positive.

You can imitate a "chart recorder" in which as you add more and more points to the right of the graph, only the most recent portions of the graph are shown. If you insert the option `fast=False`, you can pan along the whole graph when completed. You must specify the initial limits of the graph, `xmin` and `xmax`, and `xmax` must be greater than `xmin`. Listing 9.16 give a simple example. The limits will be continuously updated while scrolling.

Listing 9.16 Scrolling Graph

```
#!/usr/bin/env python3
from vpython import *
graph(scroll=True,fast=False,xmin=0,xmax=10)
g = gcurve()
for x in arange(0,30,0.1):
  rate(10)
  g.plot(x,sin(x))
```

Line 3 specifies that you want a scrolling graph, the option `Fast=Falsr` causes an option bar to appear above the plot, one of the options being the possibility to pan along the whole graph.

9.7 Exercises

Exercise 9.1 Reproduce the animation of the Pluto-Charon system, discussed in Exercise 7.5, in VPython. The radius of Pluto is approximately 1.2×10^6 m, the radius of Charon 6.06×10^5 m. Assume that the Sun light is coming from the left.

Exercise 9.2 Write a program generating a Vpython animation of the Coulomb explosion considered in Exercises 5.6, 7.6 and 7.7. Represent the explosion by an expanding sphere. Hints: If you generate only the expanding sphere, VPython automatic scaling will cause the program to display a sphere of constant radius on the canvas. We suggest to draw also two very small spheres (or any couple of very small objects) at two opposite corners of the canvas, and to set autoscaling off before starting the animation.

Exercise 9.3 Use VPython to create a 3D animation of the coupled oscillators discussed in Exercise 7.3, adding a button for stopping and restarting motion, as in Exercise 7.4. Hints: (i) taking advantage of VPython autoscaling, you can use the same units as in Exercise 7.3, so that the mathematical treatment of Listings 7.13 and 7.14 remains valid; (ii) you can represent the springs by helices of appropriate positions and lengths.

Chapter 10
Answers to the Exercises

This chapter comprises the answers to all exercises of this book. Remember that the answers to most exercises are not unique, in this case we give only one of the possible answers (possible programs that solve the exercise).

Answer 1.1 We use the formula

$$x_{1,2} = \frac{-b \mp \sqrt{b^2 - 4ac}}{2a}$$

```
>>> from math import sqrt
>>> a,b,c=2,-17,35
>>> delta=sqrt(b**2-4*a*c)
>>> x1=(-b-delta)/(2*a)
>>> x2=(-b+delta)/(2*a)
>>> print(x1,x2)
3.5 5.0
```

Answer 1.2 We can first convert the product to a string, then use the string as argument of the `len()` function.

```
>>> len(str(99999*78965321*123456499*987645))
27
```

Answer 1.3

```
>>> mylist=['Aardvark','Aargau','Abacus','Blowfish','Caper']
>>> mylist
['Aardvark', 'Aargau', 'Abacus', 'Blowfish', 'Caper']
```

Electronic supplementary material The online version of this chapter (https://doi.org/10.1007/978-3-030-45027-4_10) contains supplementary material, which is available to authorized users.

G. Moruzzi, *Essential Python for the Physicist*,
https://doi.org/10.1007/978-3-030-45027-4_10

```
>>> mylist.append('Abalone')
>>> mylist.sort()
>>> mylist
['Aardvark', 'Aargau', 'Abacus', 'Abalone', 'Blowfish', 'Caper']
```

Another possibility is

```
>>> mylist=['Aardvark','Aargau','Abacus','Blowfish','Caper']
>>> mylist.insert(3,'Abalone')
>>> mylist
['Aardvark', 'Aargau', 'Abacus', 'Abalone', 'Blowfish', 'Caper']
```

Answer 1.4

```
>>> import numpy as np
>>> a=np.array([[3.5,6,7],[6,4.5,8],[7,8,5.5]])
>>> a
array([[3.5, 6. , 7. ],
       [6. , 4.5, 8. ],
       [7. , 8. , 5.5]])
>>> b=a[:,1]
>>> b
array([6. , 4.5, 8. ])
>>> c=a[2,:]
>>> c
array([7. , 8. , 5.5])
```

Answer 1.5

```
>>> import numpy as np
>>> x=1
>>> while x<=10:
...     print('{:4.1f}␣{:10.5f}'.format(x,np.sqrt(x)))
...     x+=1
...
 1.0    1.00000
 2.0    1.41421
 3.0    1.73205
 4.0    2.00000
 5.0    2.23607
 6.0    2.44949
 7.0    2.64575
 8.0    2.82843
 9.0    3.00000
10.0    3.16228
```

Answer 1.6 The last hexadecimal digit of an integer n is a letter in the range A-F (corresponding to decimal 10–15) if the remainder of the integer division of n by 16 (decimal!) is greater than 9. Thus, one possible answer is

```
>>> n=i=0
>>> while i<10:
...     if n%16>9:
```

```
...         print('{a:3d}  {a:2X}'.format(a=n))
...         i+=1
...     n+=1
...
 10   A
 11   B
 12   C
 13   D
 14   E
 15   F
 26  1A
 27  1B
 28  1C
 29  1D
```

Answer 1.7

```
>>> A=np.array([[1,2,3],[4,5,6],[7,8,9]])
>>> B=np.array([[4,6,5],[9,8,7],[7,5,3]])
>>> C=np.dot(A,B)
>>> C
array([[ 43,  37,  28],
       [103,  94,  73],
       [163, 151, 118]])
```

Answer 1.8 The rotation matrix has the form

$$\begin{pmatrix} \cos\alpha & -\sin\alpha & 0 \\ \sin\alpha & \cos\alpha & 0 \\ 0 & 0 & 1 \end{pmatrix}$$

where alpha is the rotation angle, 20°, expressed in radians.

```
1 >>> v=np.array([1,2,3])
2 >>> alpha=np.radians(20)
3 >>> sa=np.sin(alpha)
4 >>> ca=np.cos(alpha)
5 >>> rotmat=np.array([[ca,-sa,0],[sa,ca,0],[0,0,1]])
6 >>> w=np.dot(rotmat,v)
7 >>> w
8 array([0.25565233, 2.22140538, 3.        ])
```

Line 1 creates the vector before rotation. Line 2 converts the rotation angle to radians. Lines 3 and 4 evaluate the sine and cosine of the rotation angle, and Line 5 builds the rotation matrix. At line 6, w is the vector obtained by rotating v around z by α.

Answer 1.9

```
>>> a=np.array([1,2,3,4,5,6,7,8,9])
>>> b=np.tile(a,(len(a),1))
>>> b.T*a
array([[ 1,  2,  3,  4,  5,  6,  7,  8,  9],
```

```
       [ 2,  4,  6,  8, 10, 12, 14, 16, 18],
       [ 3,  6,  9, 12, 15, 18, 21, 24, 27],
       [ 4,  8, 12, 16, 20, 24, 28, 32, 36],
       [ 5, 10, 15, 20, 25, 30, 35, 40, 45],
       [ 6, 12, 18, 24, 30, 36, 42, 48, 54],
       [ 7, 14, 21, 28, 35, 42, 49, 56, 63],
       [ 8, 16, 24, 32, 40, 48, 56, 64, 72],
       [ 9, 18, 27, 36, 45, 54, 63, 72, 81]])
```

Remember that `b.T` is the transpose of matrix `b`.

Answer 1.10

```
>>> b=np.sqrt(2)/2
>>> A=np.array([[b,-b,0],[b,b,0],[0,0,1]])
>>> eigval,eigvec=np.linalg.eig(A)
>>> eigval
array([0.70710678+0.70710678j, 0.70710678-0.70710678j, 1.+0.j])
>>> eigvec[:,0]
array([0.70710678+0.j, 0.-0.70710678j, 0.+0.j])
>>> eigvec[:,1]
array([0.70710678-0.j, 0.+0.70710678j, 0.-0.j])
>>> eigvec[:,2]
array([0.+0.j, 0.+0.j, 1.+0.j])
```

Thus there are two complex eigenvalues $E_1 = (1 + \mathrm{i})\sqrt{2}/2$ and $E_2 = (1 - \mathrm{i})\sqrt{2}/2$, and one real eigenvalue $E_3 = 1$. The corresponding eigenvectors are $\boldsymbol{v}_1 = (1, -\mathrm{i}, 0)\sqrt{2}/2$, $\boldsymbol{v}_2 = (1, \mathrm{i}, 0)\sqrt{2}/2$, and $\boldsymbol{v}_3 = (0, 0, 1)$. Matrix A is the rotation matrix by 45° around the z axis.

Answer 2.1 Script 2.11 is a possibility. The `if` statements 4–6 and 8–10 are not necessary, but their presence is instructive.

Listing 2.11 factorial.py

```
1  #!/usr/bin/env python3
2  from sys import argv
3  #
4  if len(argv)!=2:
5      print('Wrong number of arguments!')
6      exit()
7  num=int(argv[1])
8  if num<0:
9      print('{:d} is negative!'.format(num))
10     exit()
11 fact=1
12 while num>1:
13     fact*=num
14     num-=1
15 print(fact)
```

The `if` statement of Lines 4–6 exits the program via the `exit()` function if you have not typed the correct number of command-line arguments (remember that `argv[0]` is the name of the script). Line 7 converts the argument from string to integer. The `if` clause 8–10 exits the program if the input integer is negative. The `while` loop 12–14 evaluates the factorial. This is the use of the script

```
$> factorial.py 8
40320
```

Answer 2.2 Listing 2.12 is a possibility

Listing 2.12 rotate2D.py

```
1  #!/usr/bin/env python3
2  import numpy as np
3  from sys import argv
4  #
5  if len(argv)!=4:
6     print('Wrong number of arguments!')
7     exit()
8  vx=float(argv[1])
9  vy=float(argv[2])
10 alphadeg=float(argv[3])
11 #
12 alpha=np.radians(alphadeg)
13 ca=np.cos(alpha)
14 sa=np.sin(alpha)
15 RotMat=np.array([[ca,-sa],[sa,ca]])
16 vv=np.array([vx,vy])
17 ww=np.dot(RotMat,vv)
18 print(ww)
```

The `if` statement 5–7 exits the program if you have not typed the correct number of command-line arguments. Line 8–10 read the arguments and convert them to floats. Line 12 converts degrees to radians, Lines 13–15 build the rotation matrix, see Exercise 1.8. Line 17 rotates the input vector. An example follows

```
$> rotate2D.py 4 6 180
[-4. -6.]
```

Answer 2.3

Listing 2.13 FactorList.py

```
1  #!/usr/bin/env python3
2  #
3  def factorial(n):
4     fact=1
5     while n>1:
6        fact*=n
7        n-=1
8     return fact
9  for i in range(1,11):
10    print('{:2d}{:10d}'.format(i,factorial(i)))
```

Lines 3–8 define the function `factorial()`, that returns the factorial of its argument. The code is analogous to Script 2.11 of Exercise 2.1. The Loop 9–10 prints the list of the integers up to 10 and their factorials.

Answer 2.4

Listing 2.14 SqrtCbrt.py

```
#!/usr/bin/env python3
from numpy import sqrt
#
hnd=open('roots.txt','w')
hnd.write('  n  square root  cube root\n\n')
for n in range(21):
  hnd.write('{:3d}{:11.6f}{:13.6f}\n'.format(n,sqrt(n),n**(1.0/3.0)))
hnd.close()
```

Python does not offer a special function for the cubic root, so, you must use exponentiation, remembering that $\sqrt[3]{x} = x^{1/3}$. According to Sect. 1.5, the exponentiation operator is `**`. Script 2.14 generates a text file named `roots.txt`, which you can open and read with any text editor.

Answer 2.5

Listing 2.15 tart.py

```
1  #!/usr/bin/env python3
2  tria=[]
3  tria.append([])
4  tria[0].append(1)
5  n=1
6  while n<=10:
7    tria.append([])
8    tria[n].append(1)
9    j=0
10   while j<(n-1):
11     tria[n].append(tria[n-1][j]+tria[n-1][j+1])
12     j+=1
13   tria[n].append(1)
14   n+=1
15 for i in range(n):
16   print(tria[i])
```

Lines 2–4 generate the first row of the triangular array `tria`, so that `tria=[[1]]`. Each iteration of Loop 6–14 adds and fills a new row to `tria`. A new empty row is added at Line 7, resulting, at the first iteration ($n = 1$), in `tria=[[1][]]`. The first element of the new row, always a 1, is added at Line 8: `tria=[[1][1]]`. The Loop 10–12, ignored for $n = 1$, adds the intermediate elements, while the last element of the row, again a 1, is added at Line 13. Loop 15–16 prints the whole triangular array.

Answer 3.1

Listing 3.10 tanh.py

```
#!/usr/bin/env python3
import numpy as np
import matplotlib.pyplot as plt
#
xlist=np.arange(-4,4.05,0.1)
```

```
6 ylist=np.tanh(xlist)
7 plt.plot(xlist,ylist)
8 plt.show()
```

Remember that if you pass an array as argument to a `numpy` function, such as `np.tanh()`, the function operates *element-wise*. Thus Line 6 creates an array, named `ylist`, whose elements are the hyperbolic tangents of the corresponding elements of the array `xlist`.

Answer 3.2

```
>>> import matplotlib.pyplot as plt
>>> import numpy as np
>>> x=np.arange(0,4*np.pi,0.1)
>>> y=1.1+np.sin(x)
>>> plt.yscale('log')
>>> plt.plot(x,y)
>>> plt.show()
```

Answer 3.3

```
import matplotlib.pyplot as plt
import numpy as np
>>> x=np.arange(0,4*np.pi,0.1)
>>> y=np.sin(x)
>>> plt.grid=True
>>> plt.yscale('symlog',linthreshy=0.1)
>>> plt.plot(x,y)
>>> plt.show()
```

You can experiment on how changing the value of `linthreshy` affects the plot.

Answer 3.4

Listing 3.11 spiral.py

```
1 #!/usr/bin/env python3
2 import matplotlib.pyplot as plt
3 import numpy as np
4 x=[]
5 y=[]
6 for phi in np.arange(0,8*np.pi,0.1):
7     r=10*phi
8     x.append(r*np.cos(phi))
9     y.append(r*np.sin(phi))
10 plt.plot(x,y)
11 plt.show()
```

The lists of the x and y coordinates are built by Loop 6–9, which converts the polar coordinates to Cartesian coordinates.

Answer 3.5

Listing 3.12 Langevin.py

```
#!/usr/bin/env python3
```

```
2  import matplotlib.pyplot as plt
3  import numpy as np
4  #
5  def Lang(x):
6     if abs(x)>1:
7        return 1.0/np.tanh(x)-1.0/x
8     else:
9        return x/3-x**3/45      # +x**5*(2/945)
10 #
11 xlist=np.arange(-6,6.05,0.2)
12 ylist=[]
13 for x in xlist:
14    ylist.append(Lang(x))
15 plt.plot(xlist,ylist)
16 plt.show()
```

Lines 5–9 define the function `Lang(x)`. If $|x| > 1$ the function returns the actual Langevin function $L(x) = \coth(x) - 1/x$. Since $\coth(x)$ is not provided by Python, it is replaced by $1/\tanh(x)$. For $|x| \leq 1$ the function avoids the singularities of $\coth(x)$ and $1/x$ at $x = 0$ by returning the first two nonzero terms of the Taylor expansion. You can experiment what happens if you truncate the expansion at the first nonzero term, or if you include also the term in x^5, commented out in Listing 3.12.

Answer 4.1 We rewrite the system in matrix form

$$Ax = b\,, \quad \text{where} \quad A = \begin{pmatrix} 1 & 3 & -2 \\ 3 & 5 & 6 \\ 2 & 4 & 3 \end{pmatrix} \quad \text{and} \quad b = \begin{pmatrix} 5 \\ 7 \\ 8 \end{pmatrix}\,.$$

```
>>> import numpy as np
>>> A=np.array([[1,3,-2],[3,5,6],[2,4,3]])
>>> b=np.array([5,7,8])
>>> x=np.linalg.solve(A,b)
>>> x
array([-15.,    8.,    2.])
```

The solution is thus $x_1 = -15$, $x_2 = 8$ and $x_3 = 2$.

Answer 4.2 We rewrite the equation in the form

$$f(x) = \cos(2\sqrt{x}) + 1.5 - \sqrt{4 - x} = 0\,.$$

The solutions must be in the range $0 \leq x \leq 4$ for the arguments of the two square roots to be positive. We can plot the function with the methods of Chap. 3 with the following commands in interactive mode, obtaining Fig. 10.1. Figure 10.1 shows that the equation has two solutions in the interval, and their approximate locations.

```
>>> import matplotlib.pyplot as plt
>>> import numpy as np
>>> def fun(x):
```

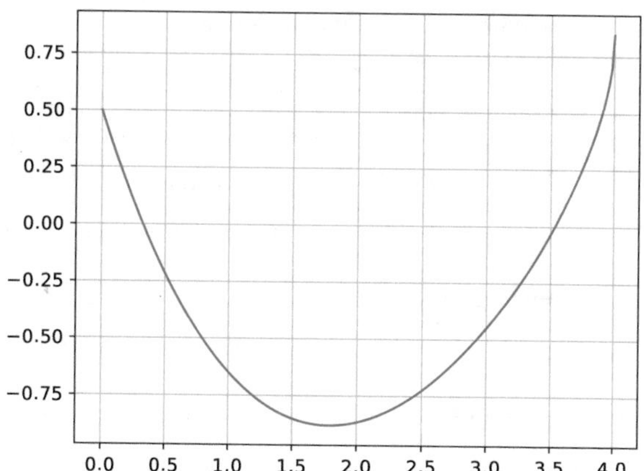

Fig. 10.1 Plot of $f(x)$

```
...     return np.cos(2.0*np.sqrt(x))-np.sqrt(4.0-x)+1.5
...
>>> x=np.arange(0,4.001,0.02)
>>> y=fun(x)
>>> plt.grid(True)
>>> plt.plot(x,y)
[<matplotlib.lines.Line2D object at 0x7f13ac11f668>]
>>> plt.show()
```

Then we can use the function `fsolve()` of the package `scipy.optimize`.

```
>>> from scipy.optimize import fsolve
>>> import numpy as np
>>> def fun(x):
...     return np.cos(2.0*np.sqrt(x))-np.sqrt(4.0-x)+1.5
...
>>> fsolve(fun,0)
array([0.32530469])
>>> fsolve(fun,3)
array([3.53166073])
```

The two numerical solutions are thus $x_1 = 0.32530469$ and $x_2 = 3.53166073$.

Answer 4.3

```
>>> from sympy import solveset
>>> from sympy.abc import x
>>> solveset(x**4-8*x**3+17*x**2+22*x-104,x)
{-2, 4, 3 - 2*I, 3 + 2*I}
```

The roots are thus $x_1 = -2$, $x_2 = 4$, $x_3 = 3 - 2\mathrm{i}$, $x_4 = 3 + 2\mathrm{i}$.

Answer 4.4 We can write and run the following script

Listing 4.9 system2b.py

```
#!/usr/bin/env python3
```

```
from scipy.optimize import fsolve
import numpy as np
#
def func(xvect):
    x,y=xvect
    r1=2*x**2-np.sin(x)+x*y-3
    r2=3*x+y**2+np.cos(x*y)-5
    return[r1,r2]
xstart=(1,1)
sol=fsolve(func,xstart)
print(''Solution:'',sol)
print(''Check:'',func(sol))
```

This is the result

```
$> System2b.py
Solution: [1.13938494 1.15148126]
Check: [-1.566746732351021e-12, -8.79296635503124e-14]
```

Thus the solutions are $x = 1.13938494$, and $y = 1.15148126$.

Answer 4.5

```
>>> from scipy.integrate import simps
>>> import numpy as np
>>> def fun(x):
...     return 1/(1+x**2)
...
>>> x=np.arange(-2,2.001,0.2)
>>> y=fun(x)
>>> simps(y,x,even='avg')
2.214293053949532
>>> x=np.arange(-2,2.001,0.1)
>>> y=fun(x)
>>> simps(y,x,even='avg')
2.214297180040977
>>> x=np.arange(-2,2.001,0.05)
>>> y=fun(x)
>>> simps(y,x,even='avg')
2.214297419595189
```

Answer 5.1 The torque of the drag force with respect to the pivot is $\tau = -\ell\beta\, v = -\ell^2\beta(\mathrm{d}\vartheta/\mathrm{d}t)$, thus (5.14) is rewritten

$$\ell^2 m\frac{\mathrm{d}^2\vartheta}{\mathrm{d}t^2} = -\ell mg\sin\theta - \ell^2\beta\,\frac{\mathrm{d}\vartheta}{\mathrm{d}t}, \quad \text{dividing by } \ell^2 m \text{ we have } \quad \frac{\mathrm{d}^2\vartheta}{\mathrm{d}t} = -\frac{\beta}{m}\frac{\mathrm{d}\vartheta}{\mathrm{d}t} - \frac{g}{l}\sin\vartheta\ .$$

We introduce the new quantity $\alpha = \beta/m = 0.2\,\mathrm{s}^{-1}$, and our second-order differential equation is equivalent to the following system of two first-order differential equations

$$\begin{aligned}\frac{\mathrm{d}\vartheta}{\mathrm{d}t} &= \omega \\ \frac{\mathrm{d}\omega}{\mathrm{d}t} &= -\alpha\omega - \frac{g}{\ell}\sin\vartheta\ .\end{aligned} \tag{5.50}$$

We cans solve (5.50) by the script

Listing 5.4 PendulumDrag.py

```
1  #!/usr/bin/env python3
2  import numpy as np
3  from scipy.integrate import odeint
4  import matplotlib.pyplot as plt
5  #
6  def dydt(y0,t, params):
7      theta,omega = y0
8      GdivL,alpha=params
9      derivs = [omega,-alpha*omega-GdivL*np.sin(theta)]
10     return derivs
11 #
12 alpha=0.2
13 GdivL=4.9
14 params=[GdivL,alpha]
15 #
16 theta0=np.pi/2.0
17 omega0=0.0
18 y0=[theta0,omega0]
19 t=np.arange(0.0,12.01,0.1)
20 sol=odeint(dydt,y0,t,args=(params,))
21 theta=sol[:,0]
22 plt.figure(figsize=(10,4))
23 plt.plot(t,theta)
24 plt.grid()
25 plt.xlabel(r'$t$/s', fontsize=22)
26 plt.ylabel(r'$\vartheta$/rad',fontsize=22)
27 plt.tight_layout()
28 plt.savefig('PlotDrag.pdf',format='pdf',dpi=1000)
29 plt.show()
```

Lines 6–10 define the function providing the derivatives to `odeint()`. The array of additional parameters `param` now includes the quantity `alpha`, i.e., $\alpha = \beta/m$. The derivatives of (5.50) are coded in Line 9. Line 12 sets the value of α, Line 14 stores g/ℓ and α into the array `params`. Line 18 stores the initial values for ϑ and ω into the array `y0`. Line 19 creates the array `t` of the times at which the pendulum position is calculated, in the range $0 \le t \le 12$ s, at intervals of 0.1 s each. Line 22 sets the ratio between the length and height of the figure. The plot is shown in Fig. 10.2.

Answer 5.2 The total charge of the sphere is $Q = Nq = 1.6 \times 10^{-11}$ C. According to Gauss's law, the electrostatic force acting on a peripheral He ion when the radius of the sphere is r will be

$$f = \frac{1}{4\pi\varepsilon_0}\frac{Qq}{r} = \frac{2.304 \times 10^{-21}}{r},$$

corresponding to an acceleration

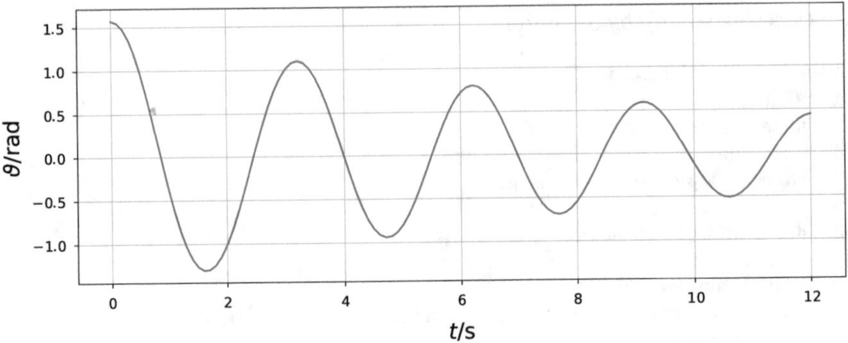

Fig. 10.2 The plot of function (4.4) between $x = -6$ and $x = 4$

$$\frac{\mathrm{d}^2 r}{\mathrm{d}t^2} = \frac{f}{m} \approx \frac{3.47 \times 10^5}{r}\,.$$

We must thus solve the following system of first-order differential equations

$$\begin{cases} \dfrac{\mathrm{d}r}{\mathrm{d}t} = v \\ \dfrac{\mathrm{d}v}{\mathrm{d}t} = \dfrac{3.47 \times 10^5}{r} \end{cases}$$

with the initial conditions $r(0) = r_0 = 10^{-7}$ m, and $v(0) = 0$ m/s. This is done by the following script, whose output is plotted in Fig. 10.3.

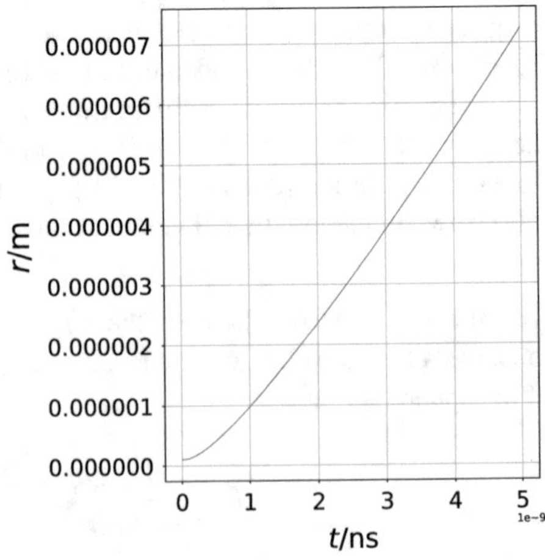

Fig. 10.3 Coulomb explosion

Listing 5.5 CoulExpl.py

```
1  #!/usr/bin/env python3
2  import numpy as np
3  from scipy.integrate import odeint
4  import matplotlib.pyplot as plt
5  #
6  const=3.47e5
7  r=1.0e-7
8  v=0.0
9  y=[r,v]
10 t=np.arange(0.0,5.01e-9,1.0e-10)
11 def dfdt(yInp,t):
12     r,vel=yInp
13     acc=const/r
14     derivs=[vel,acc]
15     return derivs
16 #
17 sol=odeint(dfdt,y,t)
18 rr=sol[:,0]
19 plt.figure(figsize=(10,10))
20 plt.plot(t,rr)
21 plt.xticks(fontsize=30)
22 plt.yticks(fontsize=30)
23 plt.grid()
24 plt.xlabel(r'$t$/ns', fontsize=40)
25 plt.ylabel(r'$r$/m',fontsize=40)
26 plt.tight_layout()
27 plt.savefig('Explosion.pdf',format='pdf',dpi=1000)
28 plt.show()
```

Variable `const`, defined at Line 6, is the quantity $Qq/(4\pi\varepsilon_0 m) = 3.47 \times 10^5$ $\mathrm{m^2\,s^{-2}}$. Function `dfst()`, defined at lines 11–15, does not need the argument `params` because `const` is used as global variable.

Answer 5.3 (a) The force acting on the proton when it is at distance r from the wire is $\boldsymbol{f} = \hat{\boldsymbol{r}}\,\lambda q/(2\pi\varepsilon_0 r)$, $\hat{\boldsymbol{r}}$ being the unit vector along the distance, pointing to the proton. Since λ is negative the force is attractive. We require the proton to follow a circular orbit of radius r_0, thus $\boldsymbol{f}$ must equal the corresponding centripetal force. If we denote the orbital velocity by v_c we have

$$\frac{|\lambda|\,q}{2\pi\varepsilon_0 r_0} = \frac{m v_\mathrm{c}^2}{r_0}\,, \quad \text{hence} \quad v_\mathrm{c} = \sqrt{\frac{|\lambda|\,q}{2\pi\varepsilon_0 m}} = 5.867 \times 10^4 \text{ m/s}\,.$$

Note that v_c is independent of the distance r_0 from the wire.

(b) The acceleration of the proton is

$$\boldsymbol{a} = \frac{\boldsymbol{f}}{m} = \hat{\boldsymbol{r}}\frac{\lambda q}{2\pi\varepsilon_0 m r} = -\hat{\boldsymbol{r}}\,\frac{K}{r}\,, \quad \text{where} \quad K = \frac{|\lambda|\,q}{2\pi\varepsilon_0 m} \simeq 3.443 \times 10^9\ \mathrm{m^2\,s^{-2}}\,. \tag{10.51}$$

Listing 5.6 is a possible program for plotting the proton trajectory. We use Cartesian coordinates with the charged wire along the z axis. We assume that at $t = 0$

the proton is located at $(x, y) \equiv (r_0, 0)$, with velocity $\boldsymbol{v} \equiv (0, v_0)$, the initial velocity being $v_0 = v_c/2$.

Listing 5.6 WireProton.py

```
1  #!/usr/bin/env python3
2  import matplotlib.pyplot as plt
3  import numpy as np
4  from scipy.integrate import odeint
5  # ........................................................ function
6  def dfdt(yInp,t):
7      x,vx,y,vy=yInp
8      r=np.sqrt(x**2+y**2)
9      alpha=np.arctan2(y,x)
10     a=-K/r
11     ax=a*np.cos(alpha)
12     ay=a*np.sin(alpha)
13     return [vx,ax,vy,ay]
14 # ................................................ numerical values
15 x=0.01              # m
16 vx=y=0.0
17 vy=2.934e4          # m/s
18 K=3.443e9           # m2/s2
19 # .................................................... input lists
20 t=np.arange(0.0,5.0e-6,1.0e-8) #1.0e-9
21 yy=[x,vx,y,vy]
22 # ..................................... solve differential equations
23 psoln=odeint(dfdt,yy,t)
24 # ........................................... plot the trajectory
25 plt.figure(figsize=(10,10))
26 plt.xticks(fontsize=14)
27 plt.yticks(fontsize=14)
28 plt.plot(psoln[:,0],psoln[:,2],'k-')
29 plt.plot(0.0,0.0,'ko')
30 plt.savefig('WireProt.pdf',format='pdf',dpi=1000)
31 plt.show()
```

Lines 6–13 define Function `dfdt` providing the derivatives to `odeint()`. Line 8 evaluates the distance `r` of the proton from the negatively charged wire. Line 9 evaluates `alpha`, the angle α that $\boldsymbol{r}$ forms with the x axis. Line 10 evaluates the magnitude of the proton acceleration, Lines 11 and 12 its x and y components. Lines 15–17 define the initial values of the proton position and velocity, Line 18 the constant `K` of (10.51). Line 20 creates the list `t` of the times at which `odeint()` must evaluate the positions and velocities of the proton. Al Line 23 Function `odeint` returns the matrix `psoln`, whose column 0 comprises the x coordinates of the proton at the required times, column 1 the x velocity components, column 2 the y positions, ... Line 25 forces the figure to have equal width and height. Line 28 plots the y coordinate of the proton, `psoln[:,2]` as a function of the x coordinate, `psoln[:,0]`. See Sect. 1.17.2 for how to slice out a column from a matrix. Line 29 plots a point representing the intersection of the charged wire with the orbit plane. The result is shown in Fig. 10.4.

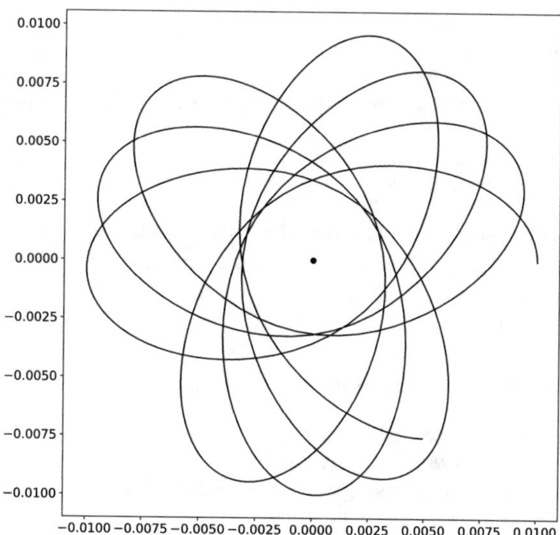

Fig. 10.4 Proton trajectory

Answer 6.1

Listing 6.5 HexaColor.py

```
#!/usr/bin/env python3
from tkinter import Tk,Canvas,E,LAST,N
import numpy as np
#
cw=400
ch=400
Ox=cw/2
Oy=ch/2
# ..................................... create root window and canvas
root=Tk()
root.title('Color Hexagon')
canvas=Canvas(root,width=cw,height=ch,background="#ffffff")
canvas.grid(row=0,column=0)
# .............................................................
colors=['red','green','blue','cyan','magenta','yellow']
# ............................................................. triang1
alpha=0.0
dAlpha=np.pi/3.0
r=200.0
i=0
while i<6:
  tr=[Ox,Oy]
  j=0
  while True:
    tr.append(Ox+r*np.cos(alpha))
    tr.append(Oy-r*np.sin(alpha))
    if j==0:
      alpha+=dAlpha
```

```
29          j+=1
30        else:
31          break
32      canvas.create_polygon(tr,fill=colors[i],outline='black')
33      i+=1
34  # ..............................................................
35  root.mainloop()
```

The hexagon is inscribed in circle of radius 200 px (`r`, defined at line 19) centered at `(Ox,Oy)`. The `while` loop of lines 21–33 generates six triangles, each with a vertex at the center of the hexagon, and each filled with one of the colors listed at Line 15.

Answer 6.2 This is a possible script:

Listing 6.6 GaussianPlot.py

```
#!/usr/bin/env python3
from tkinter import Tk,Canvas,E,LAST,N
import numpy as np
#
cw=800
ch=400
Ox=cw/2
Oy=7*ch/8
# ..................................... create root window and canvas
root=Tk()
root.title('Gaussian Plot')
canvas=Canvas(root,width=cw,height=ch,background="#ffffff")
canvas.grid(row=0,column=0)
# ........................................................... axes
ar=(20,20,5)
canvas.create_line(0,Oy,cw-1,Oy,fill='black',arrow=LAST,arrowshape=ar)
canvas.create_line(Ox,Oy,Ox,0,fill='black',arrow=LAST,arrowshape=ar)
canvas.create_text(cw-20,Oy+11,text='x',font=('Times','16','italic'))
canvas.create_text(Ox-15,15,text='y',font=('Times','16','italic'))
# ....................................................... x-ticks
dx=80
xval=-4
while xval<9:
  xpos=Ox+xval*dx
  canvas.create_line(xpos,Oy,xpos,Oy+10)
  canvas.create_text(xpos,Oy+10,text=str(xval),anchor=N)
  xval+=1
# ....................................................... y-ticks
dy=300
yval=0.2
while yval<1.01:
  ypos=Oy-yval*dy
  canvas.create_line(Ox,ypos,Ox-10,ypos)
  canvas.create_text(Ox-20,ypos,text='{:.1f}'.format(yval),anchor=E)
  yval+=0.2
# ............................... make lists of Gaussian coordinates
GaussLin=[]
```

```
w2=4.0
x=-5.0
while x<=5.0:
  y=np.exp(-(x**2/w2))
  GaussLin.append(Ox+x*dx)
  GaussLin.append(Oy-y*dy)
  x+=0.1
# ............................................... draw hyperbola
canvas.create_line(GaussLin, fill='blue')
# ...............................................................
root.mainloop()
```

Answer 7.1 This is a possibility, you are invited to experiment all possible changes.

Listing 7.11 RotaSquare1.py

```
#!/usr/bin/env python3
import numpy as np
from tkinter import *
# ........................................ initialize graphics
root=Tk()
root.title(''Rotating Square'')
# ....................................................... canvas
cw=ch=600
canvas=Canvas(root,width=cw,height=ch,background=''#ffffff'')
canvas.grid(row=0,column=0)
# ......................................................... data
Ox=ch/2
Oy=cw/2
r=200
alpha=0
delta=np.pi/2.0
dAlpha=np.radians(0.5)
corners=8*[0.5]
delay=10
# .................................................... main loop
while True:
  canvas.delete(ALL)
  canvas.create_polygon(corners,fill='red')
  canvas.update()
  for i in range(0,4):
    corners[2*i]=Ox+r*np.cos(alpha+i*delta)
    corners[2*i+1]=Oy-r*np.sin(alpha+i*delta)
    alpha+=dAlpha
    canvas.after(delay)
```

The script creates a 600×600 px canvas with white background. The rotating square is centered at the canvas center `(Ox, Oy)`. The square is inscribed in a circle of radius $r = 200$ px. One of the half-diagonals of the square forms an angle `alpha` with the x axis, `alpha` is increased by a quantity `dAlpha` at each iteration of the main loop. Quantity `delta`$= \pi/2 = 90°$ is the angle between two consecutive half-diagonals. The list `corners`$= [x_0, y_0, x_1, y_1, \ldots x_3, y_3]$ comprises the coordinates of the four corners of the square, which are reevaluated at each iteration.

Answer 7.2 Only additions or changes to Script 7.11 are commented.

Listing 7.12 RotaSquare2.py

```
#!/usr/bin/env python3
import numpy as np
from tkinter import *
# ............................................................
RunAll=True
# ..................................................... functions
def StopAll():
  global RunAll
  RunAll=False
def accel():
  global dAlpha
  dAlpha*=1.2
def decel():
  global dAlpha
  dAlpha/=1.2
# ............................................ initialize graphics
root=Tk()
root.title(''Rotating Square with Buttons'')
# ........................................................ canvas
cw=ch=600
canvas=Canvas(root,width=cw,height=ch,background=''#ffffff'')
canvas.grid(row=0,column=0)
# ....................................................... toolbar
toolbar=Frame(root)
toolbar.grid(row=0,column=1,sticky=N)
# ....................................................... buttons
nr=0
AccelButton=Button(toolbar,text=''Accelerate'',command=accel,width=11)
AccelButton.grid(row=nr,column=0,sticky=W)
nr+=1
DecelButton=Button(toolbar,text=''Decelerate'',command=decel,width=11)
DecelButton.grid(row=nr,column=0,sticky=W)
nr+=1
CloseButton=Button(toolbar,text=''Exit'',command=StopAll,width=11)
CloseButton.grid(row=nr,column=0,sticky=W)
nr+=1
# .......................................................... data
Ox=ch/2
Oy=cw/2
r=200
alpha=0
delta=np.pi/2.0
dAlpha=np.radians(0.5)
corners=8*[0.5]
delay=10
# ..................................................... main loop
while RunAll:
  if not RunAll:
    break
  canvas.delete(ALL)
  canvas.create_polygon(corners,fill='red')
```

```
52      canvas.update()
53      for i in range(0,4):
54        corners[2*i]=Ox+r*np.cos(alpha+i*delta)
55        corners[2*i+1]=Oy-r*np.sin(alpha+i*delta)
56        alpha+=dAlpha
57        canvas.after(delay)
58    canvas.destroy()
```

Line 5 defines the variable `RunAll`: the program will run as long as `RunAll` is *True*, exit otherwise. Lines 7–15 define the three functions `StopAll()`, that exits the program, `accel()`, that increases the rotation speed, and `decel()`, that decreases the rotation speed. Lines 24–25 create a toolbar at the right of the canvas. Lines 27–36 create three buttons in the toolbar, each button associated to one of the functions 7–15. The Loop 47–57 runs as long as `RunAll` is *True*.

Answer 7.3 We use as coordinates x_1, x_2 and x_3 the displacements of each sphere from its equilibrium position. Correspondingly, the elongations of the four springs are $\xi_1 = x_1$, $\xi_2 = x_2 - x_1$, $\xi_3 = x_3 - x_2$, and $\xi_4 = -x_3$, respectively. The equations of motion for the three spheres are thus written

$$\begin{cases} m\dfrac{\mathrm{d}^2x_1}{\mathrm{d}t^2} = -2kx_1 + kx_2 \\ m\dfrac{\mathrm{d}^2x_2}{\mathrm{d}t^2} = kx_1 - 2kx_2 + kx_3 \\ m\dfrac{\mathrm{d}^2x_3}{\mathrm{d}t^2} = kx_2 - 2kx_3 \end{cases} \tag{10.52}$$

Dividing the left and right side of each equation by m, and introducing the quantities $v_1 = \mathrm{d}x_1/\mathrm{d}t$, $v_2 = \mathrm{d}x_2/\mathrm{d}t$, $v_3 = \mathrm{d}x_3/\mathrm{d}t$, and $\omega^2 = k/m$ we obtain the system of first-order differential equations

$$\begin{cases} v_1 = \dfrac{\mathrm{d}x_1}{\mathrm{d}t} \\ v_2 = \dfrac{\mathrm{d}x_2}{\mathrm{d}t} \\ v_3 = \dfrac{\mathrm{d}x_3}{\mathrm{d}t} \end{cases}, \qquad \begin{cases} \dfrac{\mathrm{d}v_1}{\mathrm{d}t} = -2\,\omega^2x_1 + \omega^2x_2 \\ \dfrac{\mathrm{d}v_2}{\mathrm{d}t} = \omega^2x_1 - 2\,\omega^2x_2 + \omega^2x_3 \\ \dfrac{\mathrm{d}v_3}{\mathrm{d}t} = \omega^2x_2 - 2\,\omega^2x_3 \end{cases} \tag{10.53}$$

with the initial conditions $x_1(0) = x_0, x_2(0) = x_3(0) = v_1(0) = v_2(0) = v_3(0) = 0$. The system is solved numerically by Script 7.13.

Listing 7.13 CoupledOscill.py

```
#!/usr/bin/env python3
import numpy as np
from tkinter import *
from scipy.integrate import odeint
# ............................................ Open root window
root=Tk()
root.title('Coupled Oscillators')
```

```
8  # ........................................ add canvas to root window
9  cw=800
10 ch=400
11 canvas=Canvas(root,width=cw,height=ch,background='#ffffff')
12 canvas.grid(row=1,column=0)
13 # ..............................................................
14 O1=200
15 O2=400
16 O3=600
17 Oy=ch/2.0
18 rad=10
19 delay=20
20 omega2=1
21 # ...................................................... function
22 def f(y,t):
23     x1,v1,x2,v2,x3,v3=y
24     a1=-2*omega2*x1+omega2*x2
25     a2=omega2*x1-2*omega2*x2+omega2*x3
26     a3=omega2*x2-2*omega2*x3
27     return [v1,a1,v2,a2,v3,a3]
28 #................................................ initial values
29 x1=100
30 v1=x2=v2=x3=v3=0
31 y=[x1,v1,x2,v2,x3,v3]
32 t=[0.0,0.05]
33 # .......................................................... loop
34 while True:
35     canvas.delete(ALL)
36     canvas.create_line(0,Oy,cw,Oy, fill='black')
37     canvas.create_oval(O1+x1-rad,Oy-rad,O1+x1+rad,Oy+rad, fill='red')
38     canvas.create_oval(O2+x2-rad,Oy-rad,O2+x2+rad,Oy+rad, fill='red')
39     canvas.create_oval(O3+x3-rad,Oy-rad,O3+x3+rad,Oy+rad, fill='red')
40     canvas.update()
41     psoln = odeint(f,y,t)
42     y=psoln[1,:]
43     x1,v1,x2,v2,x3,v3=y
44     canvas.after(delay)
```

Quantities `O1`, `O2` and `O3`,defined at lines 14–16, are the equilibrium positions of the three spheres, `rad` is the common radius of the three spheres, 10 px. Quantity `omega2` is $\omega^2 = k/m = 1$ rad/s. The function `f()`, evaluating the derivatives for `odeint()`, is defined at lines 22–27. Note that no parameters are passed to the function, but `omega2` is used as *global* variable. The "real" time interval between two successive animation frames is 0.05 s, defined at Line 32.

Answer 7.4 We comment only the changes to Listing 7.13.

Listing 7.14 CouplOscillButt.py

```
#!/usr/bin/env python3
import numpy as np
from tkinter import *
from scipy.integrate import odeint
# ..............................................................
```

```
RunAll=True
RunIter=False
# ............................................ function StartStop()
def StartStop():
  global RunIter
  RunIter=not RunIter
  if RunIter:
    StartButton['text']='Stop'
  else:
    StartButton['text']='Restart'
# ............................................... function StopAll()
def StopAll():
  global RunAll
  RunAll=False
# ................................................... Open root window
root=Tk()
root.title('Coupled Oscillators with Buttons')
# ........................................ add canvas to root window
cw=800
ch=400
canvas=Canvas(root,width=cw,height=ch,background='#ffffff')
canvas.grid(row=0,column=0)
# ...................................... add toolbar to root window
toolbar=Frame(root)
toolbar.grid(row=0,column=1,sticky=N)
# .......................................... add buttons to toolbar
ButtWidth=10
StartButton=Button(toolbar,text='Start',command=StartStop,width=ButtWidth)
StartButton.grid(row=0,column=0)
CloseButton=Button(toolbar,text='Exit',command=StopAll,width=ButtWidth)
CloseButton.grid(row=1,column=0)
# ...................................................................
O1=200
O2=400
O3=600
Oy=ch/2.0
rad=10
delay=20
omega2=1
# ......................................................... function
def f(y,t):
  x1,v1,x2,v2,x3,v3=y
  a1=-2*omega2*x1+omega2*x2
  a2=omega2*x1-2*omega2*x2+omega2*x3
  a3=omega2*x2-2*omega2*x3
  return [v1,a1,v2,a2,v3,a3]
#................................................... initial values
x1=100
v1=x2=v2=x3=v3=0
y=[x1,v1,x2,v2,x3,v3]
t=[0.0,0.05]
# ............................................................. loop
while RunAll:
```

```
59        canvas.delete(ALL)
60        canvas.create_line(0,Oy,cw,Oy,fill='black')
61        canvas.create_oval(O1+x1-rad,Oy-rad,O1+x1+rad,Oy+rad,fill='red')
62        canvas.create_oval(O2+x2-rad,Oy-rad,O2+x2+rad,Oy+rad,fill='red')
63        canvas.create_oval(O3+x3-rad,Oy-rad,O3+x3+rad,Oy+rad,fill='red')
64        canvas.update()
65        if RunIter:
66            psoln = odeint(f,y,t)
67            y=psoln[1,:]
68            x1,v1,x2,v2,x3,v3=y
69        canvas.after(delay)
70    root.destroy()
```

Lines 6–7 define the global variables `RunAll` and `RunIter`. Lines 9–15 define the function `StartStop()`, that switches the value of `RunIter` between *False* and *True*. The oscillators move if `RunIter` is *True*, otherwise they are in standby. Lines 17–19 define the function `StopAll()`, that exits the program. Lines 29–36 add a toolbar with two buttons at the right of the canvas. The buttons are coupled to the functions `StartStop()` and `StopAll()`, respectively. Loop 58–69 is interrupted, and the program exited, if `RunAll` is *False*. Lines 65–68 update the positions of the three oscillators only if `RunIter` is *True*.

Answer 7.5

Listing 7.15 PlutoCharon.py

```
#!/usr/bin/env python3
import numpy as np
from tkinter import *
from scipy.integrate import odeint
# ...........................................................
root=Tk()
root.title('Pluto-Charon system')
# ................................................ open canvas
cw=ch=800
canvas=Canvas(root,width=cw,height=ch,background='white')
canvas.grid(row=0,column=0)
# ...........................................................
Ox=Oy=cw/2
# ...................................................... Pluto
m1=1.3e22                    # kg
x1=y1=0
vx1=vy1=0
rad1=10                      # px
# ..................................................... Charon
m2=1.6e21                    # kg
x2=1.9e7                     # m
y2=0
vx2=0
vy2=2e2                      # m/s
rad2=5                       # px
# .................................... gravitational constant
G=6.67408e-11
# ...........................................................
```

```
29  dt=600                    # s
30  delay=20                  # ms
31  scale=2e-5                # px/m
32  # ....................................... move to barycenter system
33  mtot=m1+m2
34  xb=(m1*x1+m2*x2)/mtot
35  yb=(m1*y1+m2*y2)/mtot
36  vxb=0
37  vyb=(m1*vy1+m2*vy2)/mtot
38  x1-=xb
39  y1-=yb
40  vx1-=vxb
41  vy1-=vyb
42  x2-=xb
43  y2-=yb
44  vx2-=vxb
45  vy2-=vyb
46  # .................................... initial conditions and times
47  y=[x1,y1,vx1,vy1,x2,y2,vx2,vy2]
48  t=[0.0,dt]
49  # ......................................................... function
50  def dfdt(yInp,t):
51    global G
52    x1,y1,vx1,vy1,x2,y2,vx2,vy2=yInp
53    distx=x2-x1
54    disty=y2-y1
55    r2=distx**2+disty**2
56    alpha=np.arctan2(disty,distx)
57    f=G/r2
58    fx=f*np.cos(alpha)
59    fy=f*np.sin(alpha)
60    ax1=fx*m2
61    ay1=fy*m2
62    ax2=-fx*m1
63    ay2=-fy*m1
64    return [vx1,vy1,ax1,ay1,vx2,vy2,ax2,ay2]
65  # ......................................................... main loop
66  while True:
67    canvas.delete(ALL)
68    canvas.create_line(0,Oy,cw,Oy,fill='black')
69    canvas.create_line(Ox,0,Ox,ch,fill='black')
70    canvas.create_oval(Ox+x1*scale-rad1,Oy-y1*scale-rad1,
71                       Ox+x1*scale+rad1,Oy-y1*scale+rad1,fill='blue')
72    canvas.create_oval(Ox+x2*scale-rad2,Oy-y2*scale-rad2,
73                       Ox+x2*scale+rad2,Oy-y2*scale+rad2,fill='red')
74    canvas.update()
75    # ....................................................
76    psoln=odeint(dfdt,y,t)
77    y=psoln[1,:]
78    x1,y1,vx1,vy1,x2,y2,vx2,vy2=y
79    canvas.after(delay)
```

Line 29 defines `dt`, the *real* time interval between two consecutive frames, as `dt=` 600 s, or 10 min. The interval between the observation of two frames on the monitor equals the computation time plus `delay`, defined as 20 ms at line 30. Quantity `scale` is the ratio of the animation size, in px, to the real size, in m. Thus, a real distance of 1.9×10^7 m will correspond to 380 px on the monitor. Lines 33–45 evaluate the positions and velocities of the two bodies in the barycentric system. Line 47 stores the initial data for the first iteration into the list `y`. Function `dfdt()`, defined at lines 50–64, evaluates the derivatives for `odeint()`. Lines 66–79 constitute the main loop of the script.

Answer 7.6

Listing 7.16 ExplosionTk.py

```
#!/usr/bin/env python3
from tkinter import *
import numpy as np
from scipy.integrate import odeint
import time
# ............................................. Global variables
RunAll=True
RunIter=False
ButtWidth=20
scale=5.0e7             # px/m
cw=ch=900               # px
cycle=20                # ms
dt=1.0e-11              # s
Ox=Oy=cw/2
# ........................................................ Origin
Ox=cw/2
Oy=ch/2
# ............................................ Start/Stop function
def StartStop():
  global RunIter
  RunIter=not RunIter
  if RunIter:
    StartButton['text']='Stop'
  else:
    StartButton['text']='Restart'
# ................................................ Exit function
def StopAll():
  global RunAll
  RunAll=False
# .................................................. root window
root=Tk()
root.title('Coulomb Explosion')
# ....................................................... canvas
canvas=Canvas(root,width=cw,height=ch,background='#ffffff')
canvas.grid(row=0,column=0)
# ...................................................... toolbar
toolbar=Frame(root)
toolbar.grid(row=0,column=1,sticky=N)
# ...................................................... buttons
nr=0
StartButton=Button(toolbar,text='Start',command=StartStop,width=11)
StartButton.grid(row=nr,column=0,sticky=N)
```

```
43  nr+=1
44  CloseButton=Button(toolbar, text='Exit', command=StopAll,width=11)
45  CloseButton.grid(row=nr,column=0,sticky=N)
46  # ....................................
47  const=3.47e5
48  r=1.0e-7           # m    initial cluster radius
49  v=0.0              # m/s  initial velocity
50  y=[r,v]            # initial values
51  t=[0,dt]
52  rr=r*scale         # px initial radius on canvasx
53  # ........................................................ dfdt
54  def dfdt(yInp,t):
55    r,vel=yInp
56    acc=const/r
57    derivs=[vel,acc]
58    return derivs
59  # ........................................ numerical time interval
60  while RunAll:
61    StartIter=time.time()
62    # ................................................ draw bodies
63    canvas.delete(ALL)
64    canvas.create_oval(Ox-rr,Oy-rr,Ox+rr,Oy+rr, fill='blue',outline='blue')
65    canvas.update()
66    # ..................................................... motion
67    if RunIter:
68      # ................................................ next step
69      psoln = odeint(dfdt,y,t)
70      y=psoln[1,:]
71      r,v=y
72      rr=r*scale
73    # ........................................... cycle duration
74    elapsed=int((time.time()-StartIter)*1000.0)
75    canvas.after(cycle-elapsed)
76  root.destroy()
```

Line 12 defines the required time interval between two successive frames on the monitor as 20 ms, while Line 13 defines the "real" time interval between two successive frames as 10^{-11} s. Lines 54–58 define the function `dfdt()`, identical to the one defined by lines 11–15 of Listing 5.5. Line 72 multiplies the "real" radius `r` of the cluster by `scale`, obtaining the radius `rr` displayed on the monitor.

Answer 7.7 Only additions to Listing 7.16 are commented. Seven lines are inserted between Lines 45 and 46

Listing 7.17 ExplosionTkLb.py

```
45  CloseButton.grid(row=nr,column=0,sticky=W)
46  nr+=1
47  # ....................................................... labels
48  TimeLab=Label(toolbar,text='{:08}␣ps'.format(0))
49  TimeLab.grid(row=nr,column=0,sticky=N)
50  nr+=1
51  RadLab=Label(toolbar,text='{:08}␣nm'.format(100))
52  RadLab.grid(row=nr,column=0,sticky=N)
```

```
53  # ........................................
54  const=3.47e5
```

The above lines locate the labels on the toolbar, below the buttons. Two lines are inserted between Lines 59 and 60 of Listing 7.6, which become Lines 67 and 68 of the present present script

```
65      return derivs
66  # ........................................ numerical time interval
67  count=0
68  tt=0.0
69  while RunAll:
```

`count is an iteration counter,` and `tt` is the "real" elapsed time. The following lines (Lines 82–88 of the present script) are inserted between Lines 72 and 73 of Script 7.6

```
81          rr=r*scale
82          tt+=dt
83      # ........................................................ labels
84      count+=1
85      if count%20==0:
86          count=0
87          TimeLab['text']='{:11.2f} ps'.format(tt*1.e12)
88          RadLab['text']='{:11.2f} nm'.format(r*1.0e9)
89      # ............................................ cycle duration
90      elapsed=int((time.time()-StartIter)*1000.0)
91      canvas.after(cycle-elapsed)
```

Line 82 increases the elapsed time if `RunIter` is *True*. Line 84 increases the iteration counter `count`. Every 20 counts the two labels in the toolbar are updated with the current values of the elapsed time in ps, and of the cluster radius in nm.

Answer 8.1

Listing 8.5 PlutoMoons0.py

```
#!/usr/bin/env python3
from tkinter import *
import numpy as np
from scipy.integrate import odeint
import time
# ............................................ Global variables
RunAll=True
RunIter=False
ButtWidth=20
scale=5.0e-6            # px/m
cw=ch=900               # px
cycle=20                # ms
dt=1200                 # s
G=6.67408e-11           # m^3/(kg s^2)
# ........................................................ Origin
Ox=cw/2
Oy=ch/2
```

```
# ........................................ Start/Stop function
def StartStop():
  global RunIter
  RunIter=not RunIter
  if RunIter:
    StartButton['text']='Stop'
  else:
    StartButton['text']='Restart'
# ................................................. Exit function
def StopAll():
  global RunAll
  RunAll=False
# ............................................... Class AstroBody
class AstroBody:
  def __init__(self,mass,radius,x,y,vx,vy,color):
    self.m=mass
    self.rad=radius
    self.x=x
    self.y=y
    self.vx=vx
    self.vy=vy
    self.col=color
    self.image=canvas.create_oval(Ox+int(scale*self.x-self.rad),\
      int(Oy-scale*self.y+self.rad),int(Ox+scale*self.x+self.rad),\
        int(Oy-scale*self.y-self.rad),fill=self.col,outline=self.col)
   # .................................................. move body
  def redraw(self):
    canvas.coords(self.image,Ox+scale*self.x-self.rad,\
      Oy-scale*self.y+self.rad,Ox+scale*self.x+self.rad,\
        Oy-scale*self.y-self.rad)
# ................................................... root window
root=Tk()
root.title('Pluto Moons')
# ......................................................... canvas
canvas=Canvas(root,width=cw,height=ch,background='#ffffff')
canvas.grid(row=0,column=0)
# ........................................................ toolbar
toolbar=Frame(root)
toolbar.grid(row=0,column=1,sticky=N)
# ........................................................ buttons
nr=0
StartButton=Button(toolbar,text='Start',command=StartStop,width=11)
StartButton.grid(row=nr,column=0,sticky=N)
nr+=1
CloseButton=Button(toolbar, text='Exit', command=StopAll,width=11)
CloseButton.grid(row=nr,column=0,sticky=N)
# ................................................. create bodies
bd=[]
bd.append(AstroBody(1.3e22,8.0,-2.137e6,0.0,0.0,-19.4,'red'))# Pluto
bd.append(AstroBody(1.6e21,6.0,1.736e7,0.0,0.0,158,'blue'))# Charon
bd.append(AstroBody(7.5e14,4,4.2656e7,0.0,0.0,151,'blue'))# Styx
bd.append(AstroBody(5.0e16,4,4.8694e7,0.0,0.0,142,'blue'))# Nix
bd.append(AstroBody(1.6e16,4,5.7783e7,0.0,0.0,130,'blue'))# Kerberos
```

```
bd.append(AstroBody(5.0e16,4.0,6.486e7,0.0,0.0,123,'blue'))# Hydra
nB=len(bd)
# .................................... coordinates and barycenter
canvas.create_line(0,Oy,cw,Oy, fill='black')
canvas.create_line(Ox,0,Ox,ch, fill='black')
# ............................................................ masses
masses=[]
for zz in bd:
  masses.append(zz.m)
# ..................................................... initial values
y=[]
for zz in bd:
  y.append(zz.x)
  y.append(zz.y)
  y.append(zz.vx)
  y.append(zz.vy)
# ......................................................... value list
t=[0.0,dt]
# .......................................................... function
def dfdt(yInp,t,mm):
  nB=len(mm)
  x=yInp[0::4]
  y=yInp[1::4]
  distx=(np.tile(x,(len(x),1))).T-x
  disty=(np.tile(y,(len(y),1))).T-y
  alpha=np.arctan2(disty,distx)
  r2=np.square(distx)+np.square(disty)
  np.fill_diagonal(r2,1.0)
  ff=np.divide(G,r2)
  np.fill_diagonal(ff,0.0)
  fx=ff*np.cos(alpha)
  fy=ff*np.sin(alpha)
  fx=(fx.T*mm).T
  fy=(fy.T*mm).T
  ax=fx.sum(axis=0)
  ay=fy.sum(axis=0)
  derivs=[0]*len(yInp)
  derivs[::4]=yInp[2::4]       # vx
  derivs[1::4]=yInp[3::4]      # vy
  derivs[2::4]=ax
  derivs[3::4]=ay
  return derivs
# ........................................ numerical time interval
while RunAll:
  StartIter=time.time()
  # ................................................ draw bodies
  for zz in bd:
    zz.redraw()
  canvas.update()
  # ...................................................... motion
  if RunIter:
    # ............................................... next step
    psoln = odeint(dfdt,y,t,args=(masses,))
```

```
124         y=psoln[1,:]
125         for i,zz in enumerate(bd):
126           zz.x=y[4*i]
127           zz.y=y[4*i+1]
128           zz.vx=y[4*i+2]
129           zz.vy=y[4*i+3]
130       # ............................................. cycle duration
131       elapsed=int((time.time()-StartIter)*1000.0)
132       canvas.after(cycle-elapsed)
133     root.destroy()
```

Lines 31–47 define the class `AstroBody`. Each instance of this class has its own mass (in kg), radius (in px), x and y coordinates (in m), v_x and v_y components of its velocity (in m/s), and color. Both `radius` and `color` do not refer to real properties of the body, but to its image on the canvas. Lines 40–42 draw the first image of the class instance on the canvas. When needed, the position of the image on the canvas is updated by the method `.redraw()` defined at lines 44–47. Lines 65–71 create a list of six instances of the class `AstroBody`, corresponding to the bodies of our problem. Lines 90–112 define the function `dfdt()`, providing the derivatives required by `odeint()`. Function `dfdt()` is equivalent to its counterpart defined at lines 217–248 of Script 8.3, the main difference being that here we are dealing with gravitational, rather than with electrostatic, forces.

Answer 8.2 Exercise 8.1 gives the initial velocities of the six bodies each with three digits after the decimal point, thus affected by rounding errors. If you evaluate the velocity of the center of mass the result is not exactly zero: it has a vertical component $v_y \approx 4.215 \times 10^{-2}$ m/s. This is why the system is slowly shifting upwards. You can correct this behavior by inserting the following lines between Lines 72 and 73 of the script.

Listing 8.6 PlutoMoons1.py

```
73  # ........................................ move to barycenter system
74  mtot=sum(zz.m for zz in bd)
75  cx=sum(zz.x*zz.m for zz in bd)/mtot
76  cy=sum(zz.y*zz.m for zz in bd)/mtot
77  cvx=sum(zz.vx*zz.m for zz in bd)/mtot
78  cvy=sum(zz.vy*zz.m for zz in bd)/mtot
79  for zz in bd:
80    zz.x-=cx
81    zz.y-=cy
82    zz.vx-=cvx
83    zz.vy-=cvy
84  # ...................................... coordinates and barycenter
```

Line 74 evaluates the total mass of the system. Lines 75–78 evaluate the x and y components of the center-of-mass position and velocity. Loop 79–83 moves everything to the barycenter system. Note that Lines 76, 77, 81 and 82 are written only for completeness: they can be omitted in the present case because all the involved y coordinates and x velocities are zero.

Answer 9.1

Listing 9.17 vPlutoCharon.py

```
#!/usr/bin/env python3
import numpy as np
from vpython import *
from scipy.integrate import odeint
# ..................................................................
scene.width=1000
scene.height=1000
scene.background=color.black
scene.title='Pluto-Charon System'
scene.lights=[]
distant_light(direction=vec(-1,0,0),color=color.white)
# ........................................... gravitational constant
G=6.67408e-11
# .......................... "real" time interval between two frames
dt=100                  # s
# ............................................ Data in Pluto's system
mp=1.3e22      # kg      Pluto
xp=0           # m
vyp=0          # m/s
mc=1.6e21      # kg      Charon
xc=1.9e7       # m
vyc=200        # m/s
# ......................................... move to barycenter system
mtot=mp+mc
xb=(mp*xp+mc*xc)/mtot
vyb=(mp*vyp+mc*vyc)/mtot
xp-=xb
vyp-=vyb
xc-=xb
vyc-=vyb
baryc=sphere(pos=vec(0,0,0),radius=1e5,color=color.yellow)
# ........................................................... Pluto
pluto=sphere(mass=mp,pos=vec(xp,0,0),radius=1.18e6,color=color.white,\
   make_trail=True,interval=2,retain=50,vel=vec(0,vyp,0))
# .......................................................... Charon
charon=sphere(mass=mc,pos=vec(xc,0,0),radius=6.06e5,color=color.white,\
   make_trail=True, interval=1,retain=100,vel=vec(0,vyc,0))
# ................................... initial conditions and times
y=[pluto.pos.x,pluto.pos.y,pluto.vel.x,pluto.vel.y,charon.pos.x,\
   charon.pos.y,charon.vel.x,charon.vel.y]
t=[0.0,dt]
# ........................................................ function
def dfdt(yInp,t):
   global G
   x1,y1,vx1,vy1,x2,y2,vx2,vy2=yInp
   distx=x2-x1
   disty=y2-y1
   r2=distx**2+disty**2
   alpha=np.arctan2(disty,distx)
   f=G/r2
   fx=f*np.cos(alpha)
```

```
52     fy=f*np.sin(alpha)
53     ax1=fx*charon.mass
54     ay1=fy*charon.mass
55     ax2=-fx*pluto.mass
56     ay2=-fy*pluto.mass
57     return [vx1,vy1,ax1,ay1,vx2,vy2,ax2,ay2]
58  # ..................................................... main loop
59  while True:
60     rate(400)
61     psoln=odeint(dfdt,y,t)
62     y=psoln[1,:]
63     pluto.pos.x,pluto.pos.y,pluto.vel.x,pluto.vel.y,charon.pos.x,\
64       charon.pos.y,charon.vel.x,charon.vel.y=y
```

Line 10 cancels the default lights, and Line 11 creates a distant white-light source coming from $x = -\infty$, simulating the Sun. Lines 17–22 give the initial positions and velocities in Pluto's reference frame, where Pluto is at rest. Lines 24–31 move to the barycentric system. Line 31 creates a small yellow sphere indicating the barycenter position.

Answer 9.2

Listing 9.18 vExplosion.py

```
#!/usr/bin/env python3
from vpython import *
import numpy as np
from scipy.integrate import odeint
scene.width=scene.height=800
scene.background=color.black
scene.title='Coulomb explosion'
# ............................................. Start/Stop function
def StartStop():
   global RunIter
   RunIter=not RunIter
   if RunIter:
      StopButton.text='Standby'.center(20)
   else:
      StopButton.text='Expand'.center(20)
# ........................................ time reversal function
def Reset():
   global r,r0,tt,v,v0,y
   r=r0
   v=v0
   tt=0.0
   y=[r,v]
# ............................................. buttons and label
StopButton=button(bind=StartStop,text='Expand'.center(20))
ResetButton=button(bind=Reset,text='Reset'.center(20))
RadiusLabel=wtext(text='{:08} nm'.format(100))
TimeLabel=wtext(text='{:11.2f} ps'.format(0))
# ............................................. Global variables
RunIter=False
dt=1.0e-11            # s
```

```
31 # ................................................ numerical values
32 const=3.47e5
33 r=r0=1.0e-7          # m     initial cluster radius
34 v=v0=0.0             # m/s   initial velocity
35 y=[r,v]              # initial values
36 t=[0,dt]
37 # ........................................................ spheres
38 cluster=sphere(pos=vec(0,0,0),radius=r,color=color.white)
39 corner1=sphere(pos=vec(-1.0e-5,-1.0e-5,0),radius=1.e-8)
40 corner2=sphere(pos=vec(1.0e-5,1.0e-5,0),radius=1.e-8)
41 scene.autoscale=False
42 # ............................................................ dfdt
43 def dfdt(yInp,t):
44   r,vel=yInp
45   acc=const/r
46   derivs=[vel,acc]
47   return derivs
48 # ....................................................... main loop
49 count=0
50 tt=0.0
51 while True:
52   rate(50)
53   # ................................................. draw cluster
54   cluster.radius=r
55   count+=1
56   if count%20==0:
57     RadiusLabel.text='{:11.2f} nm'.format(r*1.0e9)
58     TimeLabel.text='{:11.2f} ps'.format(tt*1.0e12)
59   # ................................................. Run/Standby
60   if RunIter:
61     psoln = odeint(dfdt,y,t)
62     y=psoln[1,:]
63     r,v=y
64     tt+=dt
```

Function `Reset()`, defined by Lines 17–22, restores the initial conditions of the Coulomb explosion. Lines 24 and 25 add the buttons bound to the functions `StartStop()` and `Reset()`. Lines 26 and 27 create the labels reporting the actual radius of the exploding cluster, in nm, and the elapsed time, in ps. Buttons and labels are located below the canvas.

Line 38 creates the sphere representing the exploding cluster at its initial dimensions, radius= 10^{-7} m, or 100 nm. The sphere is centered at the origin of the coordinates. Lines 39 and 40 define two small (practically invisible) spheres centered at ($x = -10^{-5}$ m, $y = -10^{-5}$ m, 0) and ($x = 10^{-5}$ m, $y = 10^{-5}$ m, 0), respectively. Because of Vpython autoscaling, these two invisible spheres will be at two opposite corners of the canvas. In the absence of Lines 39 and 40 autoscaling would cause the sphere created by Line 38 to fill most of the canvas. Line 41 switches autoscaling off. You are invited to experiment what happens if you comment out one, or more, of the lines 39–41.

Answer 9.3

Listing 9.19 vCouplOscill.py

```
#!/usr/bin/env python3
from vpython import *
from scipy.integrate import odeint
scene.width=800
scene.height=200
scene.background=color.white
# ..........................................................
RunIter=False
# ........................................ function StartStop()
def StartStop():
    global RunIter
    RunIter=not RunIter
    if RunIter:
        StopButton.text='Standby'.center(20)
    else:
        StopButton.text='Restart'.center(20)
# ..................................................... button
StopButton=button(bind=StartStop,text='Start'.center(20))
# ..........................................................
rad=30
SpringRad=15
omega2=1
# ..........................................................
Delta=200
O1=-Delta
O2=0
O3=Delta
x1=100
x2=x3=0
v1=v2=v3=0
wall1=box(pos=vec(-2*Delta,0,0),size=vec(1,200,100),color=vec(0.7,0.7,1))
wall2=box(pos=vec(2*Delta,0,0),size=vec(1,200,100),color=vec(0.7,0.7,1))
ball1=sphere(pos=vec(O1+x1,0,0),radius=rad,color=color.red)
ball2=sphere(pos=vec(O2,0,0),radius=rad,color=color.red)
ball3=sphere(pos=vec(O3,0,0),radius=rad,color=color.red)
spring1=helix(pos=vec(-400,0,0),length=Delta+x1-rad,\
  radius=SpringRad,coils=15)
spring2=helix(pos=vec(O1+x1+rad,0,0),length=Delta-x1-2*rad,\
  radius=SpringRad,coils=15)
spring3=helix(pos=vec(O2+x2+rad,0,0),length=Delta-x2-2*rad,\
  radius=SpringRad,coils=15)
spring4=helix(pos=vec(O3+x3+rad,0,0),length=Delta-x3-2*rad,\
  radius=SpringRad,coils=15)
# ..................................................... function
def f(y,t):
    x1,v1,x2,v2,x3,v3=y
    a1=-2*omega2*x1+omega2*x2
    a2=omega2*x1-2*omega2*x2+omega2*x3
    a3=omega2*x2-2*omega2*x3
    return [v1,a1,v2,a2,v3,a3]
#............................................... initial values
```

```
52  x1=100
53  v1=x2=v2=x3=v3=0
54  y=[x1,v1,x2,v2,x3,v3]
55  t=[0.0,0.05]
56  # ........................................................ loop
57  while True:
58    rate(50)
59    ball1.pos.x=O1+x1
60    ball2.pos.x=O2+x2
61    ball3.pos.x=O3+x3
62    spring1.length=Delta+x1-rad
63    spring2.pos.x=O1+x1+rad
64    spring2.length=Delta+x2-x1-2*rad
65    spring3.pos.x=O2+x2+rad
66    spring3.length=Delta+x3-x2-2*rad
67    spring4.pos.x=O3+x3+rad
68    spring4.length=Delta-x3-rad
69    if RunIter:
70      psoln=odeint(f,y,t)
71      y=psoln[1,:]
72      x1,v1,x2,v2,x3,v3=y
```

Line 20 defines the common radius of the three balls, `rad`, Line 21 the common radius of the four springs, `SpringRad`, and Line 22 the value $\omega^2 = k/m$, k being the common Hooke constant of the four springs, and m the common mass of the three balls. Line 24 defines `Delta`, the spacing between the balls when they are at their rest positions. Quantities `O1`, `O2` and `O3` are the rest positions of the three balls, `x1`, `x2` and `x3` their displacements from the rest positions, `v1`, `v2` and `v3` their velocities. Lines 31–43 create the two external walls, the three balls and the four springs at their initial positions.

Appendix A
Relevant Mathematical Functions

The math module comprises real, and a few integer, functions of real variables. The cmath module comprises complex functions of complex variables. Many functions, notably the power, logarithmic, trigonometric, ... functions have the same name in the two modules. But it is important to realize that functions of the same name belonging to different modules lead to formally different results. For instance

```
>>> from math import cos
>>> cos(0)
1.0
>>> from cmath import cos
>>> cos(0)
(1 -0j)
```

the output of the complex functions specifies that the real part of $\cos 0$ is 1, while its imaginary part is 0 (here, actually, -0). The imaginary unit, denoted by i in mathematical formulas, is coded as j in Python.

All math functions are available also in the numpy module, the important difference being that the numpy functions accept also lists or arrays as arguments. For instance:

```
>>> import numpy as np
>>> np.cos(np.pi/4)
0.7071067811865476
>>> np.cos([0,np.pi/4,np.pi/3])
array([1.        , 0.70710678, 0.5       ])
```

A.1 The math Module

The math module provides access to the mathematical functions defined by the C standard. These functions cannot be used with complex numbers as arguments and cannot return complex values: use the functions of the same name from the cmath module if you require support for complex numbers.

G. Moruzzi, *Essential Python for the Physicist*,
https://doi.org/10.1007/978-3-030-45027-4

In the following we list the functions provided by this module. Except when explicitly noted otherwise, all return values are floats. Symbol *NaN*, standing for *Not a Number*, is a numeric data type value representing an undefined or unrepresentable value, especially in floating-point calculations, for instance, the division 0/0.

A.1.1 Number-Theoretic and Representation Functions

- `math.ceil(x)` Returns the smallest *integer* greater than or equal to `x`, $\lceil x \rceil$.
- `math.copysign(x,y)` Returns `x` with the sign of `y`.
- `math.fabs(x)` Returns $|x|$, the absolute value of `x`.
- `math.factorial(x)` Returns $x!$, the factorial of `x`.
- `math.floor(x)` Returns $\lfloor x \rfloor$, the largest *integer* less than or equal to `x`.
- `math.fmod(x,y)` Returns x mod y, the remainder when `x` is divided by `y`. In other words: the *dividend* x and the *divisor* y are assumed real, the *quotient* is assumed integer, and the function returns the *remainder*.
- `math.frexp(x)` Returns the mantissa and exponent of `x` as the pair `(m,y)`, where `m` is a float $0.5 \leq m < 1$, `y` is an integer, and $x = m \times 2^y$.
- `math.fsum(iterable)` Returns an accurate floating point sum of values in the list `iterable`, avoiding the loss of precision due to multiple roundings in the intermediate partial sums:

```
>>> sum([.1,.1,.1,.1,.1,.1,.1,.1,.1,.1])
0.9999999999999999
>>> math.fsum([.1,.1,.1,.1,.1,.1,.1,.1,.1,.1])
1.0
```

- `math.isfinite(x)` Returns *True* if `x` is neither an infinity nor a *NaN* (Not a Number).
- `math.isinf(x)` Returns *True* if `x` is a positive or negative infinity.
- `math.isnan(x)` Returns *True* if `x` is a *NaN*.
- `math.ldexp(x,y)` Returns $x \times 2^y$. This is essentially the inverse of function `frexp()`.
- `math.modf(x)` Returns the fractional and integer parts of `x` as the pair `(f,n)`, where $x = n + f,\ 0 \leq f < 1$, and n is an *integer*.
- `math.trunc(x)` Returns the truncated integer value of `x`. Note that `trunc()` returns a *float*, while `floor()` returns an *integer*.

A.1.2 Power and Logarithmic Functions

- `math.exp(x)` Returns e^x.
- `math.expm1(x)` Returns $e^x - 1$.

- `math.log(x[,base])` Returns the logarithm of x to the specified base (defaults to e).

```
>>> print(math.log(10),math.log(10,10))
2.302585092994046 1.0
```

- `math.log1p(x)` Returns $\ln(1+x)$, the natural logarithm of $(1+x)$.
- `math.log2(x)` Returns $\log_2 x$, equivalent to `math.log(x,2)`.
- `math.log10(x)` Returns $\text{Log}\, x = \log_{10} x$, equivalent to `math.log(x,10)`.
- `math.pow(x,y)` Returns x^y.
- `math.sqrt(x)` Returns $\sqrt{x}$.

A.1.3 *Trigonometric Functions*

- `math.acos(x)` Returns $\arccos x$.
- `math.asin(x)` Returns $\arcsin x$.
- `math.atan(x)` Returns $\arctan x$.
- `math.atan2(y,x)` Returns $\arctan(y/x)$, in radians. The result is between $-\pi$ and π. Since the signs of both arguments are known, `atan2()` computes the correct quadrant.
- `math.cos(x)` Returns $\cos x$.
- `math.hypot(x,y)` Returns the Euclidean norm, $\sqrt{x^2+y^2}$.
- `math.sin(x)` Returns $\sin x$.
- `math.tan(x)` Returns $\tan x$.

A.1.4 *Angular Conversion*

- `math.degrees(x)` Converts angle x from radians to degrees.
- `math.radians(x)` Converts angle x from degrees to radians.

A.1.5 *Hyperbolic Functions*

- `math.acosh(x)` Returns $\text{arcosh}\, x$.
- `math.asinh(x)` Returns $\text{arsinh}\, x$.
- `math.atanh(x)` Returns $\text{artanh}\, x$.
- `math.cosh(x)` Returns $\cosh x$.
- `math.sinh(x)` Returns $\sinh x$.
- `math.tanh(x)` Returns $\tanh x$.

Note that the prefix "arc" in the trigonometric functions is the abbreviation for *arcus*, while the prefix "ar" in the hyperbolic functions stands for *area*.

A.1.6 Special Functions

- `math.erf(x)` Returns the error function

$$\mathrm{erf}(x) = \frac{1}{\sqrt{\pi}} \int_{-x}^{x} \mathrm{e}^{-t^2}\,\mathrm{d}t = \frac{2}{\sqrt{\pi}} \int_{0}^{x} \mathrm{e}^{-t^2}\,\mathrm{d}t \tag{A.1}$$

- `math.erfc(x)` Returns the complementary error function $1 - \mathrm{erf}(x)$

$$\mathrm{erfc}(x) = 1 - \mathrm{erf}(x) = \frac{2}{\sqrt{\pi}} \int_{x}^{\infty} \mathrm{e}^{-t^2}\,\mathrm{d}t \tag{A.2}$$

- `math.gamma(x)` Returns $\Gamma(x)$, the Gamma function.
- `math.lgamma(x)` Returns $\ln(|\Gamma(x)|)$, the natural logarithm of the absolute value of the Gamma function at x.

A.1.7 Constants

- `math.pi` Mathematical constant, the ratio of circumference of a circle to its diameter, $\pi = 3.14159\ldots$
- `math.e` Mathematical constant, base of the natural logarithms, $\mathrm{e} = 2.71828\ldots$

A.2 The `cmath` Module

A complex number is written as `x+yj`, where `x` is the real part, `y` the imaginary part, and `j` the coding of the imaginary unit i.

```
z=5.5+3.2j
>>> print(z)
(5.5+3.2j)
>>> print(z.real)
5.5
>>> print(z.imag)
3.2
```

A.2.1 *Conversions to and from Polar coordinates*

- `cmath.polar(z)` Returns the representation of `z` in polar coordinates, i.e., `(r,phi)` where `r` is the modulus of `z` and `phi` is the phase of `z`, thus $z = r\,e^{i\varphi}$. Function polar(z) is equivalent to `(abs(z), phase(z))`.
- `cmath.rect(r,phi)` Returns the complex number `z` with polar coordinates `r` and `phi`.

```
>>> import cmath
>>> z=5+4j
>>> cmath.polar(z)
(6.4031242374328485, 0.6747409422235526)
>>> cmath.rect(6.4031242374328485, 0.6747409422235526)
(5+4j)
>>>
```

A.2.2 *Power and Logarithmic Functions*

- `cmath.exp(z)` Returns e^z.
- `cmath.log(z[,base])` Returns the logarithm of `z` to the given base. If the base is not specified, returns the natural logarithm of `z`. There is one branch cut, from 0 along the negative real axis to $-\infty$, continuous from above.
- `cmath.log10(z)` Returns $\text{Log}z = \log_{10} z$. This has the same branch cut as `cmath.log()`.
- `cmath.sqrt(z)` Returns $\sqrt{z}$. This has the same branch cut as `cmath.log()`.

A.2.3 *Trigonometric Functions*

- `cmath.acos(z)` Returns $\arccos z$x. There are two branch cuts: one extends right from 1 along the real axis to ∞, continuous from below. The other extends left from -1 along the real axis to $-\infty$, continuous from above.
- `cmath.asin(z)` Returns $\arcsin z$. This has the same branch cuts as `cmath.acos()`.
- `cmath.atan(z)` Returns $\arctan z$. There are two branch cuts: One extends from `1j` along the imaginary axis to ∞j, continuous from the right. The other extends from $-1j$ along the imaginary axis to $-\infty j$, continuous from the left.
- `cmath.cos(z)` Returns $\cos z$.
- `cmath.sin(z)` Return the $\sin z$.
- `cmath.tan(z)` Return the $\tan z$.

A.2.4 Hyperbolic Functions

- `cmath.acosh(z)` Returns $\operatorname{arcosh} z$. There is one branch cut, extending left from 1 along the real axis to $-\infty$, continuous from above.
- `cmath.asinh(z)` Returns $\operatorname{arsinh} z$. There are two branch cuts: One extends from 1j along the imaginary axis to ∞j, continuous from the right. The other extends from $-1j$ along the imaginary axis to $-\infty j$, continuous from the left.
- `cmath.atanh(z)` Returns $\operatorname{artanh} z$. There are two branch cuts: One extends from 1 along the real axis to ∞, continuous from below. The other extends from -1 along the real axis to $-\infty$, continuous from above.
- `cmath.cosh(z)` Returns $\cosh z$.
- `cmath.sinh(z)` Returns $\sinh z$.
- `cmath.tanh(x)` Returns $\tanh z$.

A.2.5 Classification Functions

- `cmath.isinf(z)` Returns *True* if the real or the imaginary part of z is $+\infty$ or $-\infty$.
- `cmath.isnan(z)` Returns *True* if the real or the imaginary part of z is not a number (*NaN*).

A.2.6 Constants

- `cmath.pi` A complex number whose real part is π, and whose imaginary part is 0.
- `cmath.e` A complex number whose real part is e, and whose imaginary part is 0.

Appendix B
Building Your Own Library

B.1 Writing a Module

As a program gets longer, it is convenient to split it into several files for easier maintenance. You might also want to use one, or more, of your functions in several separate programs without copying their definitions into each program. You can do this by putting the definitions of the functions that you plan to use often into a file, which Python calls a *module*. Definitions from a module can then be imported into other modules, into a script, or into interactive mode with the usual `import` command.

Thus, a module is a file containing Python definitions and statements. The file name is the module name with the suffix `.py` appended. Within a module, the module's name (as a string) is available as the value of the global variable `__name__`. For instance, you can use your favorite text editor to create the file of Listing B.1, comprising two function definitions. If you call the file `mymath.py`, the module name will be `mymath`.

Listing B.1 mymath.py
```
import math
# ............................................ factorize
def factorize(n):
  sqnf=int(math.ceil(math.sqrt(float(n))))
  factors = []
  while n%2==0:
    factors.append(2)
    n=n//2
  i=3
  while i<=sqnf:
    while n%i==0:
      factors.append(i)
      n=n//i
      sqnf=int(math.ceil(math.sqrt(float(n))))
    i+=2
  if n!=1:
    factors.append(n)
  return factors
```


G. Moruzzi, *Essential Python for the Physicist*,
https://doi.org/10.1007/978-3-030-45027-4

```
19 # ........................................... FiboSeq
20 def FiboSeq(n):    # return Fibonacci sequence up to n
21   result = []
22   a,b = 0, 1
23   while b < n:
24     result.append(b)
25     a, b = b, a+b
26   return result
```

Line 1 imports the module `math`, needed by the function `factorize()`. Python modules can import one another, here our module imports one of Python built-in modules. Lines 3–18 define the function `factorize()`, the same that we already met in Script 2.2. Lines 20–21 define the simple function `FiboSeq()`, that returns the Fibonacci sequence up to n, the argument of the function. As you know, a Fibonacci sequence is a sequence of integers, whose first two elements are 0 and 1, and each successive element is the sum of the two elements that precede it.

If you are working in the same directory where you stored your `mymath.py` file, you can import your `mymath` module without problems. In interactive mode you can simply type

```
>>> import mymath
>>> mymath.factorize(3333333333333331)
 [31, 1499, 717324094199]
```

or, alternatively,

```
>>> from mymath import FiboSeq
>>> FiboSeq(400)
 [0, 1, 1, 2, 3, 5, 8, 13, 21, 34, 55, 89, 144, 233,
     377]
```

You can also write a script accessing your module. Let us call it, for instance, `TestLib.py`,

Listing B.2 TestLib.py

```
#!/usr/bin/env python3
from mymath import factorize
fact=factorize(3333333333333331)
print(fact)
```

which runs as follows

```
$> TestLib.py
 [31, 1499, 717324094199]
```

Thus, everything is fine as long as you are working in the same directory where you stored your module. However, it is often convenient to build a separate directory where to store all your modules (all your library). Let us call this directory, for instance, `~/python/lib`, where the tilde (~) stands for your home directory, which is platform dependent, namely

- `/home/<username>/` under Linux;
- `/Users/<username>` under macOS;

- `C:\Users\<username>`
  under Microsoft Windows Vista.

Unfortunately, if you are working in a different directory your module is no longer automatically accessible. If you try to import it, this is how Python reacts

```
>>> import mymath
Traceback (most recent call last):
  File "<stdin>", line 1, in <module>
ModuleNotFoundError: No module named 'mymath'
```

But don't panic: there are two possibilities to import the modules from your new directory to wherever you wish in your computer:

1. **Temporary access**: when you ask to import a module named, for instance, `mymath`, the interpreter first searches for a *built-in* module with that name. If not found, it then searches for a file named `mymath.py` in a list of directories contained in the built-in variable `sys.path`. Variable `sys.path` is a list of directories comprising (i) the directory containing the input script (or the current directory); (ii) the directories listed in the *environment variable* `PYTHONPATH`, if existent, see Sect. B.2; and iii) the installation-dependent default. This is how you can add your new directory to the list of `sys.path` when you are in interactive mode

   ```
   >>> import sys
   >>> sys.path.append('/home/<username>/python/lib')
   ```

   remember that the argument of `.append()` is operating-system dependent. From now on your module is accessible. But if you exit and reenter Python interactive mode the accessibility is lost, unless you change again the value of `sys.path` by retyping the two lines. If you wish to access your module from inside a Python script, these must be the first script lines:

   ```
   \#!/usr/bin/env python3
   import sys
   sys.path.append('/home/<username>/python/lib')
   ```

   The second and third lines must be added to every script that calls your module, but otherwise this "change" is permanent.
2. **Permanent access**: You can define an appropriate *environment variable*, as discussed in Sect. B.2. This will make your modules accessible from everywhere in your computer and forever.

B.2 The PYTHONPATH Environment Variable

Environment variables are *global system variables* accessible by all the processes running in a computer. Environment variables are used to store system-wide values such as the directories, or lists of directories, where to search for, for instance, executable programs or libraries. In other words, environment variables tell programs from what directories to read files, in what directories to write files, where to store temporary files …

The environment variable PYTHONPATH sets the search path for importing *your* python modules. Thus, for instance, you can store your mymath module, and all your other modules, in a directory called ~/python/lib. Obviously, you can name the directory as you wish, and locate is wherever it is most convenient for you in your file system. By default, the variable PYTHONPATH is not defined on your computer, but you can easily add it. The procedure is not difficult, but operating-system dependent:

- Under Linux, look for the hidden file .bashrc in your home directory, edit it with your favorite editor and add the following two lines at the end

  ```
  PYTHONPATH=:.:~/python/lib
  export PYTHONPATH
  ```

  the right-hand-side of the first line defines two fields, separated by a colon (":"), where Python should search for your modules: the dot (" .") stands for the current directory, ~/python/lib is the directory where you write your modules. Under Linux and macOS, the character tilde (~) is a shorthand for /home/<username>/. Changes to your search path will take effect when you start a new shell.
- Under macOS, edit the file .profile and add the two lines

  ```
  PYTHONPATH = ${PYTHONPATH}:~/python/lib
  export PYTHONPATH
  ```

  Again, changes to your search path will take effect when you start a new shell.
- Under Windows, execute the following steps

  1. Access "System Settings" from your Control Panel.
  2. Click on the "Advanced" tab.
  3. Click on the "Environmental Variables" button on the bottom of the screen.
  4. Click the "New" button under the "System Variables" section.
  5. Type PYTHONPATH in the "Variable" field. Type the path for Python modules in the value field. Click "OK" when you are finished setting the PYTHONPATH environmental variable.

After doing this, your modules are permanently accessible from any of your directories.

Appendix C
The *integrate* Subpackage

C.1 simps()

The function

```
simps(y,x=None,dx=1,axis=-1,even='avg')
```

integrates $y(x)$ in dx using samples along the x axis and the composite Simpson's rule. If x is *None*, a uniform spacing equal to dx is assumed. An even number n of samples implies an odd number $n - 1$ of subintervals. Since Simpson's rule requires an even number of subintervals, the parameter even controls how the extra subinterval is handled. This is how the function arguments are handlesd.

- y is the array of the function values to be integrated.
- x is the optional array of the abscissae of the y values. If x is not given, a uniform spacing equal to dx is assumed.
- dx , optional, is the spacing between the integration points along the x axis. It is only used when x is *None*. The default value 1.
- axis int, optional, forget this.
- even optional, can be 'avg', 'first' or last'. Active only if the number n of y values is even, implying an odd number $n - 1$ of subintervals. Value 'first' means using Simpson's rule for the first $n - 2$ subintervals and the trapezoidal rule on the last subinterval, see Sects. 4.7.2 and 4.7.3. Value 'last' means using Simpson's rule for the last $n - 2$ subintervals and the trapezoidal rule on the first subinterval. Value 'avg' means averaging the results of 'first' and 'last'.

G. Moruzzi, *Essential Python for the Physicist*,
https://doi.org/10.1007/978-3-030-45027-4

C.2 odeint()

The function

```
odeint(func,y0,t,args=(),Dfun=None,col_deriv=0,
    full_output=0,ml=None,mu=None,
  rtol=None,atol=None,tcrit=None,h0=0.0,hmax=0.0,
     hmin=0.0,ixpr=0,mxstep=0,
  mxhnil=0,mxordn=12,mxords=5,printmessg=0)
```

integrates a system of ordinary differential equations using the solver *lsoda* from the FORTRAN library *odepack*. This is the argument list of the function, only the first three arguments are mandatory, and used in the present book:

- `func(y,t0,...)` is a function that computes the derivative(s) of the function(s) `y` at `t0`. It can actually point to an array of functions.
- `y0` is an array comprising the initial conditions on `y`
- `t` is an array comprising the time points at which to solve for `y`. The initial value point should be the first element of the array
- `args` is an optional tuple comprising the extra arguments needed for evaluating the function.
- `Dfun` is the gradient (Jacobian) of `func`. Not needed here.
- `col_deriv` is an optional Boolean. It must be set to `True` if `Dfun` defines derivatives down columns (faster), otherwise `Dfun` should define derivatives across rows.
- `full_output` is an optional Boolean, `True` if to return a dictionary of optional outputs as the second output
- `printmessg` is an optional Boolean, `True` if the convergence message must be printed

Appendix D
Methods for Drawing and Writing on the Canvas

This appendix lists some relevant methods for drawing geometrical shapes and writing text on the Tkinter canvas.

- `create_arc(x0, y0, x1, y1, dash=, dashoffset=, disableddash=, disabledfill=, disabledoutline=. disabledoutlinestipple=, disabledstipple=, disabledwidth=, extent=, fill=, offset=, outline=, outlinestipple=, start=, state=, stipple=, style=, tags=, width=)`
  Draws an arc from the ellipse inscribed in the rectangle with opposite vertices (x_0, y_0) and (x_1, y_1). The arc is delimited by the angles (in degrees) `start` and `extent`.
- `create_line(coord, activedash, activefill, activestipple, activewidth, arrow=, arrowshape=, capstyle=, dash=, dashoffset=, disableddash=, disabledfill=, disabledstipple=, disabledwidth=, joinstyle=, splinesteps=, state=, stipple=, tags=, width=)`
  Draws a polyline comprising n segments, whose vertices (x_0, y_0), (x_1, y_1), … (x_n, y_n) are elements of the list `coord`.
- `create_oval(x0, y0, x1, y1, activedash=, activefill=, activeoutline=, activeoutlinestipple=, activestipple=, activewidth=, dash=, dashoffset=, disableddash=, disabledfill=, disabledoutline=, disabledoutlinestipple=, disabledstipple=, disabledwidth=, fill=, offset=, outline=, outlineoffset=, outlinestipple=, smooth=, splinesteps=, state=, stipple=, tags=, width=)`
  Draws an ellipse inscribed in the rectangle with opposite vertices (x_0, y_0) and (x_1, y_1).

G. Moruzzi, *Essential Python for the Physicist*,
https://doi.org/10.1007/978-3-030-45027-4

- `create_polygon(coord, activedash=, activefill=, activeoutline=, activewidth=, dash=, dashoffset=, disableddash=, disabledfill=, disabledoutline=, disabledoutlinestipple=, disabledstipple=, disabledwidth=, fill=, joinstyle=, offset=, outline=, outlineoffset=, outlinestipple=, smooth=, splinesteps=, state=,stipple=, tags=, width=)`
  Draws a polygon on n sides and n vertices, the vertex coordinates (x_0, y_0), (x_1, y_1), $\ldots(x_{n-1}, y_{n-1})$ being the $2n$ elements of the list `coord`.
- `create_rectangle(x0, y0, x1, y1, activedash=, activefill=, activeoutline=, activeoutlinstipple=, activestipple=, activewidth=, dash=, dashoffset=, disableddash=, disabledfill=, disabledoutline=, disabledoutlinestipple=, disabledstipple=, disabledwidth=, fill=, offest=, outline=, outlineoffset=, outlinestipple=, state=, stipple=, tags=, width=)`
  Draws a rectangle with opposite vertices (x_0, y_0) and (x_1, y_1).
- `create_text(x, y, activefill=, activestipple=, anchor=, disabledfill=, disabledstipple=,fill=, font=, justify=, offest=, state=, stipple=, text=, width=)`
  Writes text at position (x, y). By default the text is centered on this position.

Meaning of the mandatory function arguments

- `coord`: A list of the form $[x_0, y_0, x_1, y_1, \ldots x_n, y_n]$ where (x_i, y_i) are the coordinates of the vertices of the polyline in the case of `create_line()`, and of the vertices of the polygon in the case of `create_polygon()`.
- `x, y`: Position of the text for `create_text()`.
- `x0, y0, x1, y1`: Opposite vertices of the rectangle for `create_rectangle()`. Opposite vertices for the rectangle bounding the ellipse for `create_arc()` and `create_oval()`.

Meaning of the optional function arguments

- `activedash, activefill, activeoutline, activeoutlinstipple, activestipple, activewidth`: These options specify the `dash`, `fill`, `stipple`, and `width` values to be used when the object is active, that is, when the mouse is over it.
- `anchor`: the default is `anchor=tk.CENTER`, meaning that the text is centered vertically and horizontally around the position (x, y). The other options are (tk.NW), (tk.N), (tk.NE), (tk.E), (tk.SE), (tk.S), (tk.SW), (tk.W), corresponding to the compass points.
- `arrow` The default is for the line to have no arrowheads. Use `arrow=tk.FIRST` to get an arrowhead at the (x_0, y_0) end of the line. Use `arrow=tk.LAST` to get an arrowhead at the far end. Use `arrow=tk.BOTH` for arrowheads at both ends.

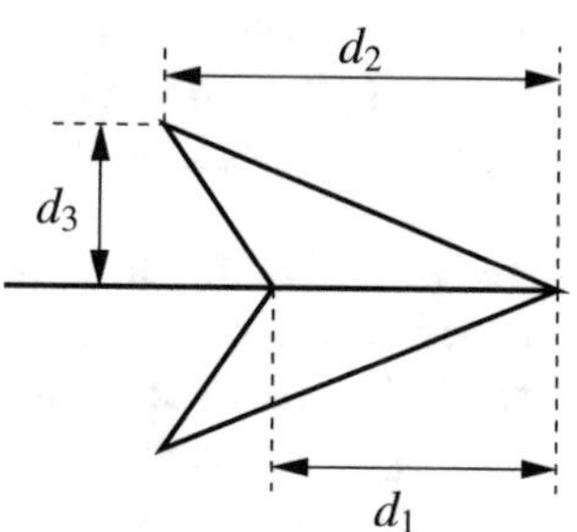

Fig. D.1 Shape of the arrowhead

- `arrowshape`: A tuple (d_1, d_2, d_3) describing the shape of the arrowheads added by the arrow option. The meaning of the three quantities d_1, d_2 and d_3 is shown in Fig. D.1. Thus, in general, the arrowhead is a quadrilateral, more specifically a *kite*, either convex or concave. It is a convex kite if $d_2 < d_1$: a *rhombus* in the special case $d_1 = 2d_2$. It is a concave kite, also called *dart*, if $d_2 > d_1$, as in the case of Fig. D.1. In the special case $d_2 = d_1$ it is an isosceles triangle of base $2d_3$, which can be considered a degenerate kite. Thus the command `canvas.create_line(0, 50, 100, 50, arrow=LAST, arrowshape=(20 ,20, 5))` creates a horizontal line from $(0, 50)$ to $(100, 50)$ with an isosceles triangle as arrowhead. The arrowhead starts at $x = 100 - 20 = 80$. The base of the triangle is $2 \times d_3 = 10$. All measures are in pixels. The default values for `arrowshape` are $d_1 = 8$, $d_2 = 10$, $d_3 = 3$.
- `capstyle`: Specifies the shape of the ends of the line. The options are (i) `tk.BUTT`: the end of the line is cut off square at a line that passes through the endpoint; (ii) `tk.PROJECTING`: the end of the line is cut off square, but the cut line projects past the endpoint a distance equal to half the line's width; (iii) `tk.ROUND`: the end describes a semicircle centered on the endpoint.
- `dash`: Must be a tuple of integers. The first integer specifies how many pixels should be drawn. The second integer specifies how many pixels should be skipped before starting to draw again, and so on. When all the integers in the tuple are exhausted, they are reused in the same order until the border is complete. The default is a solid line.
- `dashoffset`: If you specify a dash pattern, the default is to start the specified pattern at the beginning of the line. The `dashoffset` option allows you to specify that the start of the dash pattern occurs at a given distance after the start of the line
- `disableddash, disabledfill, disabledstipple, disabledwidth`: The `dash`, `fill`, `stipple`, and `width` values to be used when the item is in the `tk.DISABLED` state.
- `extent`: width of the arc in degrees. The arc starts at the angle given by the `start` option and extends counterclockwise for `extent` degrees.
- `fill`: the color to use in drawing the line. Default is `fill='black'`.

- `font`: if you don't like the default font, set this option to any font value. Examples: `font=('Helvetica', '16')` for a 16-point Helvetica regular; `font=('Times', '24', 'bold italic')` for a 24-point Times bold italic.
- `joinstyle`: For lines that are made up of more than one line segment, this option controls the appearance of the junction between segments. The options are (i) `tk.ROUND`: the join is a circle centered on the point where the adjacent line segments meet; (ii) `tk.BEVEL`: a flat facet is drawn at an angle intermediate between the angles of the adjacent lines; (iii) `tk.MITER`: the edges of the adjacent line segments are continued to meet at a sharp point. The default style is `tk.ROUND`
- `justfy`: for multi-line text displays, this option controls how the lines are justified: `tk.LEFT` (the default), `tk.CENTER`, or `tk.RIGHT`.
- `offset`: the purpose of this option is to match the stippling pattern of a stippled line with those of adjacent objects.
- `outline`: the color of the border around the outside of the geometric shape, for `create_arc()`, `create_oval()`, `create_polygon()` and `create_rectangle()`. Default is black.
- `outlinestipple`: if the outline option is used, this option specifies a bitmap used to stipple the border. Default is black, and that default can be specified by setting `outlinestipple=''`.
- `smooth`: if `True`, the line is drawn as a series of parabolic splines fitting the point set. Default is `False`, which renders the line as a set of straight segments.
- `splinesteps`: if the `smooth` option is `true`, each spline is rendered as a number of straight line segments. The `splinesteps` option specifies the number of segments used to approximate each section of the line; the default is `splinesteps=12`.
- `start`: Starting angle for the slice, in degrees, measured from $+x$ direction. If omitted, you get the entire ellipse.
- `state`: normally, line items are created in state `tk.NORMAL`. Set this option to `tk.HIDDEN` to make the line invisible; set it to `tk.DISABLED` to make it unresponsive to the mouse.
- `stipple`: to draw a stippled line, set this option to a bitmap that specifies the stippling pattern, such as `stipple='gray25'`. See Sect. 5.7, "Bitmaps" for the possible values.
- `style`: the default is to draw the whole arc; use style=`tk.PIESLICE` for this style. To draw only the circular arc at the edge of the slice, use `style=tk.ARC`. To draw the circular arc and the chord (a straight line connecting the endpoints of the arc), use `style=tk.CHORD`.
- `tags`: if a single string, the line is tagged with that string. Use a tuple of strings to tag the line with multiple tags.
- `text`: the text to be displayed in the object, as a string. Use newline characters (`'\n'`) to force line breaks.
- `width`: the line's width (`create_line()`), or width of the border of the geometric shape (`create_arc()`, `creat_oval()`, `create_polygon()` and

`create_rectangle()`). Default value is 1 pixel. In the case of `create_text()`, if you don't specify a width option, the text will be set inside a rectangle as long as the longest line. However, you can also set the width option to a dimension, and each line of the text will be broken into shorter lines, if necessary, or even broken within words, to fit within the specified width.

Appendix E
Unicode Characters

We advise to use LaTeXwhenever possible when subscripts, superscripts, Greek letters, … are needed in text superposed to Python plots. However, this is not always possible, in particular when writing text on Tkinter buttons and labels. Often Unicode symbols offer an acceptable alternative. UTF-8 is a variable width character encoding capable of encoding all 1 112 064 Unicode characters (they include, for instance, Greek, Hebrew, Chinese, Japanese, Korean …characters) using one to four bytes. UTF-8 is fully backward-compatible with ASCII encoding. The Tables E.1 and E.2 give the two-byte encodings for the numerical subscripts and superscripts, and for the Greek letters.

For instance, x_0 is rendered by `x\u2080`, and y^2 by `y\u2072`.

Each two-byte code is represented by a backslash (\) followed by a `u` and four hexadecimal digits.

Table E.1 UTF-8 numerical superscripts and subscripts

Number	0	1	2	3	4	5	6
Superscript	\u2070	\u2071	\u2072	\u2073	\u2074	\u2075	\u2076
Subscript	\u2080	\u2081	\u2082	\u2083	\u2084	\u2085	\u2086
Number	7	8	9				
Superscript	\u2077	\u2078	\u2079				
Subscript	\u2087	\u2088	\u2089				

G. Moruzzi, *Essential Python for the Physicist*,
https://doi.org/10.1007/978-3-030-45027-4

Table E.2 UTF-8 greek letters

Α	\u0391	Ι	\u0399	Ρ	\u03A1	α	\u03B1	ι	\u03B9	ρ	\u03C1
Β	\u0392	Κ	\u039A	Σ	\u03A2	β	\u03B2	κ	\u03BA	σ	\u03C3
Γ	\u0393	Λ	\u039B	Τ	\u03A3	γ	\u03B3	λ	\u03BB	τ	\u03C4
Δ	\u0394	Μ	\u039C	ϒ	\u03A4	δ	\u03B4	μ	\u03BC	υ	\u03C5
Ε	\u0395	Ν	\u039D	Φ	\u03A5	ε	\u03B5	ν	\u03BD	φ	\u03C6
Ζ	\u0396	Ξ	\u039E	Χ	\u03A6	ζ	\u03B6	ξ	\u03BE	χ	\u03C7
Η	\u0397	Ο	\u039F	Ψ	\u03A7	η	\u03B7	ο	\u03BF	ψ	\u03C8
Θ	\u0398	Π	\u03A0	Ω	\u03A9	ϑ	\u03B8	π	\u03C0	ω	\u03C9

Appendix F
Tkinter Events

Tkinter events can be key presses or mouse operations by the user. For each widget (root window, canvas, frame ...), it is possible to bind Python functions and methods to an event. The syntax is widget.bind(event, handler)
where widget can be the root window itself (usually called root in the programs of this book) or one of its "children", like a canvas or a frame, event is a keyboard or mouse event, and handler is usually a user-defined function. See Sect. 7.4 for some first examples.

A description of the mouse and keyboard events is given in the two following sections.

F.1 Mouse Events

A mouse event is generated whenever the mouse moves, a mouse button is clicked or released, or the mouse wheel is turned. Table F.1 lists the mouse events and their string codings. The x and y coordinates of the mouse-cursor location when a mouse event occurs are passed to the handler function through the argument as event.x and event.y. See Listing 7.4 for examples.

F.2 Keyboard Events

A keyboard event occurs when a keyboard key is pressed or released. Events corresponding to pressing printable keys like, for instance, a, A, or 8 are simply represented by the strings '<a>', '<A>' or '<8>'. The release of the same keys is represented by the strings '<KeyRelease-a>', '<KeyRelease-A>', and '<KeyRelease-8>'. The string '<Key>' binds to any key press, which key was actually pressed is passed to the handler function through the argument as event.char. Binding to events like pressing Ctrl plus another key,

G. Moruzzi, *Essential Python for the Physicist*,
https://doi.org/10.1007/978-3-030-45027-4

Table F.1 String codings of mouse events

String	Event
`'<Button-`*n*`>'`	where *n* can be 1, 2, 3, 4 or 5. A mouse button is pressed over the widget. Button 1 is the leftmost button, button 2 is the middle button (where available), button 3 the rightmost button. Buttons 4 and 5 refer to turning the mouse wheel forward and backward, respectively
`'<ButtonPress-`*n*`>'`	synonym of `'<Button-`*n*`>'`
`'<B`*n*`-Motion>'`	The mouse is moved, with mouse button *n* being held down
`'<ButtonRelease-`*n*`>'`	Button *n* was released
`'<Double-Button-`*n*`>'`	Button *n* was double clicked. You can use Double or Triple as prefixes
`'<Enter>'`	The mouse pointer entered the widget (this event does not mean that the user pressed the `<Enter>` key!)
`'<Leave>'`	The mouse pointer left the widget

Table F.2 String codings of keyboard events: keypad and non-printable keys

String	Key	String	Key
`'<Alt_L>'`	Left-hand Alt	`'<KP_Down>'`	Down arrow on the keypad
`'<Alt_R>'`	Right-hand Alt	`'<KP_End>'`	End on the keypad
`'<BackSpace>'`	Backspace	`'<KP_Enter>'`	Enter on the keypad
`'<Cancel>'`	Del	`'<KP_Home>'`	Home on the keypad
`'<Caps_Lock>'`	CapsLock	`'<KP_Insert>'`	Ins on the keypad
`'<Control_L>'`	Left-hand Ctrl	`'<KP_Left>'`	Left arrow on the keypad
`'<Control_R>'`	Right-hand Ctrl	`'<KP_Multiply>'`	`*` on the keypad
`'<Delete>'`	Del	`'<KP_Next>'`	PageDown on the keypad
`'<Down>'`	Down arrow	`'<KP_Prior>'`	PageUp on the keypad
`'<End>'`	End	`'<KP_Right>'`	Right arrow on the keypad
`'<Escape>'`	Esc	`'<KP_Subtract>'`	− on the keypad
`'<F1>'`	Function key F1	`'<KP_Up>'`	up arrow on the keypad
…	…	`'<Next>'`	PageDown
`'<F12>'`	Function key F12	`'<Num_Lock>'`	NumLock
`'<Home>'`	Home	`'<Pause>'`	Pause
`'<Insert>'`	Ins	`'<Print>'`	PrintScrn
`'<Left>'`	Left arrow	`'<Prior>'`	PageUp
`'<KP_0>'`	0 on the keypad	`'<Return>'`	Enter
…	…	`'<Right>'`	Right arrow
`'<KP_9>'`	9 on the keypad	`'<Scroll_Lock>'`	ScrollLock
`'<KP_Add>'`	+ on the keypad	`'<Shift_L>'`	Left-hand Shift
`'<KP_Decimal>'`	`.` on the keypad	`'<Shift_R>'`	Right-hand shift key
`'<KP_Delete>'`	Del on the keypad	`'<Tab>'`	Tab
`'<KP_Divide>'`	/ on the keypad	`'<Up>'`	Up arrow

Table F.3 String codings of keyboard events: some special printable keys

String	Key	String	Key	String	Key
'<backslash>'	\	'<bar>'	|	'<quotedbl>'	"
'<dollar>'	$	'<percent>'	%	'<ampersand>'	&
'<slash>'	/	'<parenleft>'	(	'<parenright>'	)
'<equal>'	=	'<question>'	?	'<minus>'	–
'<at>'	@	'<asterisk>'	*	'<greater>'	>
'<less>'	<	'<comma>'	,	'<semicolon>'	;
'<period>'	.	'<colon>'	:	'<underscore>'	_
'<plus>'	+	'<minus>'	–	'<sterling>'	£

like, for instance Ctrl+a, is obtained through strings like `'<Control-a>'`. In Listing 8.3 Lines 149 and 150 bind the key combinations `'<Control-plus>'` and `'<Control-minus>'` to the handler functions `ScaleUp()` and `ScaleDown()`, respectively.

Table F.2 shows the representations of the non-printable keys and of the keys of the numeric keypad, the representations of these latter always starting with `KP_`. Some printable keys are not represented by their symbols on the keyboard, but have special representations. A few of them are reported in Table F.3.

Appendix G
VPython

G.1 Colors

As we saw in Sect. 9.3.1, VPython defines colors according to the RGB additive model, specifying the intensities of the three primary colors red, green and blue. Each intensity is specified by a real number in the closed interval [0, 1]. Thus, when in Listing 9.2 we write `color=vec(0.9,0.9,0.2)` we specify a color where red and yellow have 90% of their maximum possible intensity, while blue has 20% of its possible maximum intensity. VPython also offers predefined color names, listed in Table G.1 with their RGB equivalents. Thus, writing `color=vector(1,1,0)` is equivalent to writing `color=color.yellow`.

G.2 Extrusion: Shapes and Paths Library

The *shapes library* helps in creating complex 2D shapes by creating and combining basic geometric shapes. Originally a shape is created in the xy plane. These shapes are then used for extrusions along given paths. The *paths library* is similar, in the sense that it comprises the same geometric figures as the shapes library, but the perimeters of the geometric figures are used as extrusion paths rather than for delimiting the

Table G.1 Color names

Name	RGB code	Name	RGB code
`color.red`	`vector(1,0,0)`	`color.orange`	`vector(1,0.6,0)`
`color.green`	`vector(0,1,0)`	`color.cyan`	`vector(0,1,1)`
`color.blue`	`vector(0,0,1)`	`color.magenta`	`vector(1,0,1)`
`color.purple`	`vector(0.4,0.2,0.6)`	`color.black`	`vector(0,0,0)`
`color.yellow`	`vector(1,1,0)`	`color.white`	`vector(1,1,1)`

G. Moruzzi, *Essential Python for the Physicist*,
https://doi.org/10.1007/978-3-030-45027-4

cross section. In using the paths library, it is helpful to think of a path as being originally created in the xy plane. The path is then tipped back onto the xz plane, with the top of the path pointing in the $-z$ direction (into the computer monitor).

A list of the available geometric figures follow.

G.2.1 Rectangle

This command creates a 2D rectangular shape. The command

```
MyRect=shapes.rectangle(width=10,height=6)
```

is equivalent to the command

```
MyRect=[[5,-3],[5,3],[-5, 3],[-5, -3],[5, -3]]
```

The shape can be visualized in VPython by executing the following command, which extrudes the rectangular shape into the screen, along the line from `vec(0,0,0)` to `vec(0,0,-3)`:

```
MyExtr=extrusion(shape=MyRect,path=[vector(0,0,0),
    vector(0,0,-3)],\
  color=vector(1,0.5,0))
```

the result is shown in Fig. G.1. You can check that what you created is actually a cuboid by changing the direction of its axis. This is done by the further command

```
MyExtr.axis=vector(1,1,1)
```

which generates Fig. G.2. A rotated rectangle can be obtained like this:

```
MyRect=shapes.rectangle(width=10,height=6,rotate=pi/6)
```

This produces a rectangle rotated counterclockwise by $\pi/6$ rad (30°). A negative angle rotates the figure clockwise.

It is possible to define a hollow rectangle, i.e., a rectangle with a rectangular hole, by using the thickness parameter:

```
MyRect=shapes.rectangle(width=10,height=6,thickness
    =0.1)
MyExtr=extrusion(shape=MyRect,path=[vector(0,0,0),
    vector(0,0,3)],\
  color=vector(1,0.5,0))
MyExtr.axis=vector(1,1,1)
```

This defines a "rectangular frame" with a size 10×6, and a thickness of 0.1 times the shortest side of the rectangle, the height in this case. This results in a rectangle with a rectangular hole. The extrusion is shown in Fig. G.3.

Fig. G.1 Rectangle

Fig. G.2 Extruded rectangle

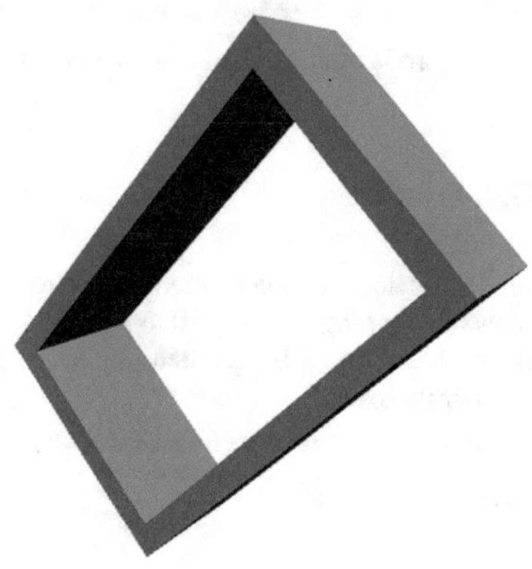

Fig. G.3 Extruded frame

G.2.2 Circle

A circle of radius 2, centered at the origin, is created by the command

```
MyCirc=shapes.circle(radius=2)
```

By default, the circle is approximated by a regular polygon of 64 sides. You can change this by specifying a value for the attribute `np`. For instance, the command

```
MyCirc=shapes.circle(radius=2,np=8)
```

actually generates an octagon.

An annulus is obtained if the `thickness` attribute is set to a positive value. If the thickness is 0.1, the actual thickness will be 0.1 times the radius. See Sect. G.2.1 above.

You can generate a circular sector by specifying the start and the final angle, corresponding to attributes `angle1` and `angle2`. The angles are measured in radians counterclockwise from the x axis.

```
MySlice=shapes.circle(radius=2,angle1=0,angle2=pi/2)
```

A circular sector can have a `thickness`.

G.2.3 Ellipse

An ellipse centered at ($x = 0$, $y = 0$), of horizontal minor axis 1 and vertical major axis 1.5, is created by the command

```
MyEllipse=shapes.ellipse(width=1,height=1.5)
```

If `height` is omitted it is set to be equal to the `width`, and the output is a circle.

G.2.4 Arc

The following command creates a circular arc centered at `pos=(0,0)`, with radius=2, ranging from $\vartheta = 0$ to $\vartheta = \pi/4$ (45°)) counterclockwise from the x axis. If no `thickness` is specified, the arc is given a very small thickness so that its contour is closed.

```
MyArc=shapes.arc(radius=2,angle1=0,angle2=pi/4)
```

G.2.5 Line

The following command creates a straight line going from ($x = 1$, $y = 0$) to ($x = 1$, $y = 1$), divided into 20 equal line segments. If no thickness is specified, the line is given a very small thickness so that the contour is closed. The defaults are `start=(0,0)`, `end=(0,1)`.

```
MyLine=shapes.line(start=(1,0),end=(1,1),np=20)
```

G.2.6 Triangle

The following command creates a triangle object centered at ($x = 0$, $y = 0$), with all sides equal to 5. A triangular frame is obtained if the `thickness` attribute is set to a positive value.

```
MyTriangle=shapes.triangle(length=5)
```

G.2.7 Pentagon, Hexagon, Octagon, Ngon

Analogous to `triangle`. In the case of `ngon` you must also specify the number of sides. An heptagon is generated by the command

```
MyHeptagon=shapes.ngon(np=7,length=5)
```

G.2.8 Star

The command

```
MyStar=shapes.star(n=5,iradius=1,radius=3)
```

creates a star object centered at ($x = 0$, $y = 0$), with 5 beams sticking out. The convex vertices lie on a circle of radius 3 and the concave vertices on a concentric circle of radius 1. The default value of `radius` is 1, the defalult value of `iradius` is `0.5*radius`.

G.2.9 Trapezoid

The command

```
MyTrapezoid=shapes.trapezoid(pos=[-2,3],width=5,
    height=1,top=3)
```

creates an isosceles trapezoid object with its center at $x = -2$, $y = 3$), with `width=5` (the width of the base), `height=1`, and `top=3`. If `top` is omitted, it is set to be equal to half of the `width`. The trapezoid can be rotated and rounded as well as scaled. A trapezoidal frame is obtained if the `thickness` attribute is set to a positive value. If the thickness is 0.1, the actual thickness will be 0.1 times the length of the shortest side.

Index

G. Moruzzi, *Essential Python for the Physicist*,
https://doi.org/10.1007/978-3-030-45027-4

Printed in the United States
by Baker & Taylor Publisher Services